Infrared Astronomy – Seeing the Heat

from William Herschel to the Herschel Space Observatory

Infrared Astronomy – Seeing the Heat

from William Herschel to the Herschel Space Observatory

David L. Clements

Imperial College London, UK

CRC Press
Taylor & Francis Group
Boca Raton London New York

CRC Press is an imprint of the
Taylor & Francis Group, an **informa** business

CRC Press
Taylor & Francis Group
6000 Broken Sound Parkway NW, Suite 300
Boca Raton, FL 33487-2742

First issued in hardback 2019

© 2015 by Taylor & Francis Group, LLC
CRC Press is an imprint of Taylor & Francis Group, an Informa business

No claim to original U.S. Government works

ISBN-13: 978-1-4822-3727-6 (hbk)

Library of Congress Cataloging-in-Publication Data

Clements, David L., 1964-
   Infrared astronomy : seeing the heat : from William Herschel to the Herschel Space Observatory / David L. Clements.
      pages cm
   "A CRC title."
   Includes bibliographical references and index.
   ISBN 978-1-4822-3727-6 (hardcover : alk. paper) 1. Infrared astronomy. 2. Astronomy--History. I. Title.

QB470.C54 2015
522'.683--dc23
2014039071

Visit the Taylor & Francis Web site at
http://www.taylorandfrancis.com

and the CRC Press Web site at
http://www.crcpress.com

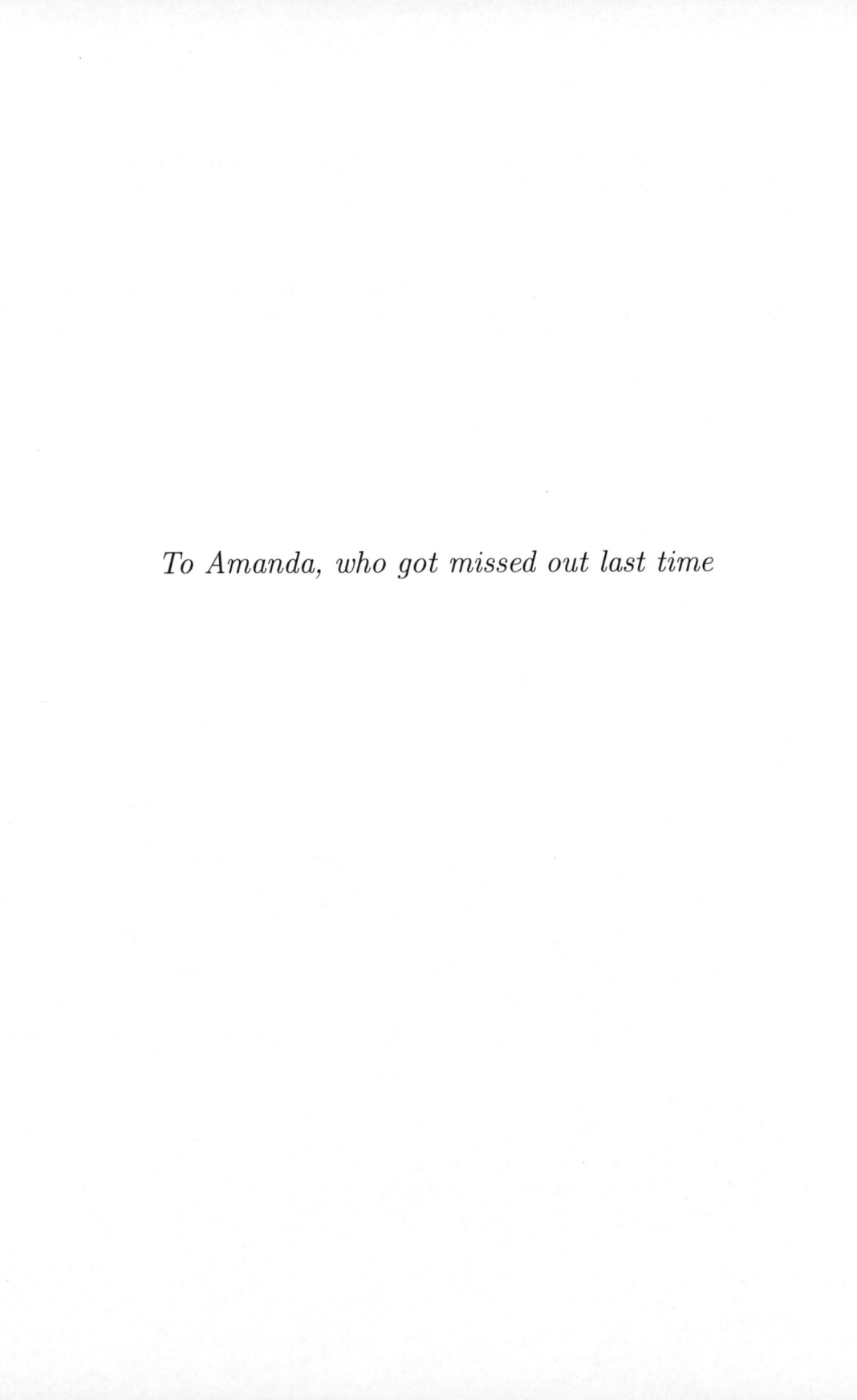

To Amanda, who got missed out last time

The preamble to Chapter 6 previously appeared in a different form in lablit.com and is used here by kind permission of the editor, Jenny Rohn.

The preamble to Chapter 7 previously appeared in a different form in the *Rocket Science* anthology, published by Mutation Press, and is used here by kind permission of the editor, Ian Sales.

Contents

List of Figures

Foreword: The Arp 220 Moment

I am at the James Clerk Maxwell Telescope (JCMT) on top of the extinct volcano Mauna Kea in Hawaii. It is about 4 am, and the observing night is beginning to draw to a close. This is my first solo observing run. There's nobody at the telescope to help me other than the telescope operator who isn't familiar with the science I'm trying to do.

The weather is great, with clear skies and excellent conditions for the submillimetre wavelengths I'm using. And yet my targets have not cooperated.

I've detected none of the galaxies I was meant to be observing and am beginning to get a little desperate. Everything is working properly, and I can clearly see the pen of the paper plotter swing from left to right as the telescope observes one of our calibration sources. But my targets remain elusive.

It's time to look at something else, since weather like this should not be wasted.

I have to get the next set of observations ready. Instead of returning to another of my elusive galaxies I ask the operator to point the telescope at Arp220, a galaxy known to be very bright in the far-infrared but which still hasn't been observed by the JCMT, which only opened a few years before these observations.

A few minutes later the building has finished rotating and has pointed the huge dish of the telescope at my target. The observations begin, and the pen on the chart recorder starts to swing, first one way, as we look at Arp220, and then the other, as we look away.

I have a detection.

I think about this for a few moments. Gradually the significance dawns on me.

Arp220 is a galaxy about 250 million light years away from Earth. The submillimetre light I'm detecting has been travelling towards the Earth for 250 million years. When it was emitted, there were only primitive reptiles and insects on the surface of this planet. Dinosaurs and mammals had yet to evolve. Primates and hominids were a distant figment of the future's imagination. The island I am standing on wasn't here, and the configuration of the continents was vastly different.

This ancient light has been spreading across the universe for a quarter of a billion years, until now, when some vanishingly small fraction of it has been captured by the JCMT, and has reached down the electric cables of the telescope to push the graph plotter's pen from side to side.

I'm hooked.

From this day on I know I want to remain an observational astronomer.

Preface

When I told my old PhD supervisor, Prof Bob Joseph, now based at the Institute for Astronomy in Hawaii, that I was writing a book on infrared astronomy, his first reaction was to say, 'Good for you!'. His second reaction was to say, 'You could have done that 25 years ago, but you can't do that today'.

He was right.

Twenty five years ago, infrared astronomy was its own little discipline, as were radio astronomy, X-ray astronomy, optical astronomy and others. Among observational astronomers and cosmologists, wavelength chauvinism was the rule. X-ray astronomers looked at X-ray results and, to a large extent, didn't place as much weight on results derived by astronomers working at other wavelengths. The same applied to everyone else.

But change was in the air. At first this seemed to be a new specialism of its own, multiwavelength astronomy, the idea of which was that you looked at your chosen targets with all the tools available.

Today things are very different. We are all, to a greater or lesser extent, mutiwavelength astronomers, using all the tools and wavelengths available to us to understand the universe. While I might be driven by the infrared, selecting sources of interest from surveys done, for example, by the Herschel Space Observatory, I'm quite happy to use optical, radio and X-ray data to better understand my targets. The same goes for astronomers whose sources are selected in the X-rays, optical or radio. This means that infrared astronomy is now a tool for everyone to use, and all of astronomy is, to some extent, related to what you can see in the infrared.

Because of this, the book you have in front of you is something of a Trojan Horse. While it will certainly show a lot of new work in astronomy that relies on observations in the infrared, what it is really doing is telling you about how the universe works, ranging in scale from planets in our own Solar System, to the universe as a whole. That, after all, is what astronomers and astrophysicists are trying to find out, no matter what observational tools we might happen to use.

It must also be said that the universe is a big place, and there is a lot to be said about it. Sadly, limitations of space and time mean that I haven't been able to include everything that we have learned about the universe, or even everything that we have learned from the infrared. At the end of this book there is a list of suggested further reading, so if you want to learn more these resources should be your next step. I have also tried to provide references to academic publications, so those interested in seeing more of the nitty gritty of what astronomers get up to can look up the original source material. Finally, I have also set up a web page on my blog which provides additional material for the book, including things like movies which are rather hard to present on the pages of a book. This web page can be found at `http://davecl.wordpress.com/infrared-astronomy-book`. My blog, at `http://davecl.wordpress.com`, includes coverage of recent astronomical results as well as many other things. I can also be found on twitter as `@davecl42`.

No book is written in isolation. The resources I have drawn on include scientific publications that date back several centuries, the excellent archives of astronomical images provided by NASA, ESA, ESO and many other observatories, and the expertise and experience of my many friends and colleagues in the field. My thanks to you all - this really would not have been possible without you. In particular I must thank those academic colleagues who have read various chapters of this book to make sure I haven't made any serious mistakes, or who have helped me when I've had questions. These heroes are Dr Brian O'Halloran, Dr Yvonne Unruh, Dr Stephen Feeney, Dr Emily Drabek-Maunder, Dr Subhanjoy Mohanty, Prof Andrew Jaffe, Dr Tom Geballe, Prof Steve Warren, Dr Nathalie Skrzypek and Dr Gören Pilbratt. Getting the science right isn't the only thing necessary in a book, and so I must also thank my writing group, the London Catherd, for many years of help with my writing, and specifically Alys Sterling and Nick Moulton who went beyond the call of duty in reading and commenting on the whole thing. I must also thank my editor, Francesca McGowan, who thought the random idea I had for a book as I cycled to work one day was worth pursuing, and who has helped me through the delicate process of producing my first book. The Taylor & Francis LaTeX gurus Marcus Fontaine and Shashi Kumar also rode to the rescue when they were needed, and the proofreading abilities of Michele Dimont were essential.

Any errors that remain, scientific, spelling, grammatical, typesetting or otherwise, remain my own responsibility, not theirs.

Dr David L. Clements, Imperial College London, May 2014

Author

David L. Clements is a Senior Lecturer in the Physics Department of Imperial College London. He has a degree in physics and a PhD in astrophysics from Imperial College. He has worked at Oxford University, The European Southern Observatory Headquarters, near Munich, the Institut d'Astrophysique Spatiale, near Paris, and at Cardiff University. His research is in the broad areas of extragalactic astronomy and observational cosmology, specialising in studies of dust in galaxies and the role that dusty galaxies play in galaxy formation and evolution. He has worked on the Herschel and Planck missions as Manager of the London Planck Analysis Centre, and head of the Herschel Data Processing and Science Analysis Software Centre London. As an infrared astronomer he has used data from all the major infrared satellite missions, from IRAS to Herschel, and many ground based telescopes. He is the author of over 200 scientific papers and also writes science fiction stories and popular science articles, which have been published in *Analog*, *Nature Futures* and *Astronomy Now* among others. His blog can be found at `http://davecl.wordpress.com`, which also hosts material associated with this book, and on twitter at `@davecl42`. This is his first book.

CHAPTER 1

Finding the Heat

1.1 REDDER THAN RED

On a sunny day in the year 1800, William Herschel is working in a darkened room, lit only by a slit of sunlight passing in through the window. This light passes through a prism which splits the light into a rainbow of colours, spreading them out from violet to red across the desk where he is conducting an experiment.

During previous experiments, Herschel observed that the heating power of the Sun, as measured by the thermometers arrayed across his desk, varied with colour. Red light from the Sun caused the temperature of the thermometer placed in that part of the spectrum to rise faster than those exposed to blue or violet light. Today's experiment, though, is going to be different. Today, Herschel is going to explore beyond what he can see with his own eyes, to see if something in the light from the Sun produces radiant heating effects while remaining undetectable to the human eye.

For the current experiment, Herschel has mounted three thermometers on a wooden block (Figure 1.1), with the central thermometer exposed to the spectrum of sunlight, and those beside it, out of the sunlight, able to record temperature variations in the room. Any heating experienced by the central thermometer and not by the others will be due to the light of the Sun, whether this can seen by Herschel's eyes or not.

He tests the apparatus by repeating the work he has already done in the blue and red parts of the spectrum, confirming results he obtained a month before. Then he moves the central thermometer out of the light he can see, and into the part of the spectrum where light redder than red would be seen, if it were detectable by the human eye.

Surprisingly, the temperature of the thermometer continues to rise.

He moves the instrument further into the red, recording the heating strength of this invisible light.

After completing his observations, Herschel finds that the maximum heating effect occurs half an inch beyond the visible part of the spectrum [79], and

Figure 1.1 William Herschel discovering infrared light. (Courtesy of NASA.)

that there is still heating an inch and a half beyond the spectrum he can see, which itself is eight inches long, from violet to red.

William Herschel has discovered infrared light.

1.2 THE ELECTROMAGNETIC SPECTRUM

Astronomy relies on signals from space which, until very recently, have all been carried by electromagnetic waves, whether in the form of visible light, infrared light, radio waves or gamma rays. To understand infrared astronomy, and the reason why it plays a central role in our understanding of the universe, we have to first look at the nature of electromagnetic radiation and the physical processes that produce infrared light.

Herschel's discovery of the infrared was clearly a necessary first step towards infrared astronomy, but it also opened up broader questions about the nature of light. These questions would echo around the scientific world for the rest of the 19th century and would lead, in the early days of the 20th century, to the discovery of quantum mechanics.

At the time of Herschel's discovery scientists were still debating whether light was a wave phenomenon or, as Newton had thought, whether it was made up of discrete particles. We shall see later that light actually has the properties of both waves and particles, but the dominant view in the 19th century was that light was a wave phenomenon. This means that every kind of light has an associated wavelength.

When Herschel shone sunlight through his prism, the glass bent, or refracted, the incoming light rays at different angles depending on their wavelength, with bluer shorter wavelength light being diverted more than redder, longer wavelength light. This split the incoming white light, made up of many different wavelengths, into the spectrum of colours that Herschel could see, with the infrared light he couldn't see, but could measure with his thermometer, bent even less by the prism.

The electromagnetic spectrum covers a far wider range of wavelengths than just the part of the spectrum that was analysed by Herschel, see Figure 1.2. Very long wavelengths correspond to radio emission, while very short wavelengths correspond to X-ray and gamma-ray radiation. The difference in wavelengths of different types of electromagnetic waves determine how they interact with matter. Bluer visible light, for example, is refracted more strongly by glass than redder light. This allows glass prisms to produce spectra from visible light. At shorter, ultraviolet wavelengths, glass becomes opaque, so it can't be used at these wavelengths. The atmosphere itself is opaque to a wide variety of wavelengths, including ultraviolet and X-rays, which is useful since it means that it protects us from such radiation which would otherwise be quite harmful.

Infrared light lies at wavelengths between the end of the visible, at around one micron (one millionth of a metre, about half the size of a particle of

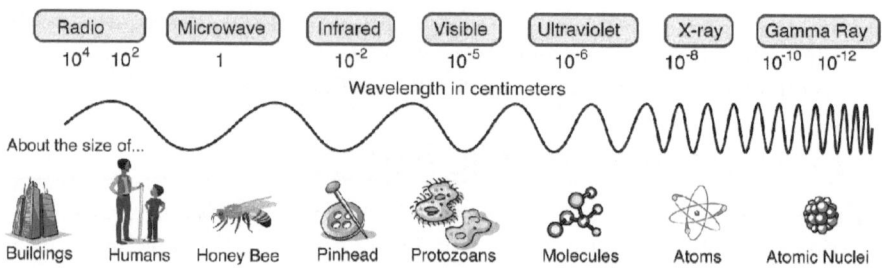

Figure 1.2 The electromagnetic spectrum.
A comparison of the wavelengths of different parts of the electromagnetic spectrum to various physical bodies of the same size, from atoms to buildings. (Courtesy of NASA.)

cigarette smoke) to 1 millimetre[1]. Within this broad range of wavelengths, covering a part of the electromagnetic spectrum that is roughly a thousand times as wide as the optical, there are three sub-ranges called the near-infrared (about 1 - 5 microns in wavelength), the mid-infrared (about 5 - 25 microns in wavelength) and the far-infrared (about 25 - 500 microns in wavelength). There is also a borderland between the infrared and the radio called the sub-millimetre. This covers the spectrum from 500 microns in wavelength to a few millimetres. Submillimetre astronomy will also be covered in this book where it is relevant, and where the processes involved are broadly similar to those going on at shorter, far-infrared wavelengths.

It turns out that the 'infrared' light that Herschel detected beyond the red end of the visible spectrum that day in 1800 is not what we would call infrared light in the context of astronomy today. It is now regarded as an extended part of the visible spectrum, often called the optical, since we can apply exactly the same techniques to it that are used for the part of the electromagnetic spectrum that we can see. Just to add to the confusion, this light, with a wavelength of about 0.8 microns [145], is still regarded as infrared light by photographers.

1.3 THERMAL RADIATION

What makes the light that we see, and that we observe with telescopes?

There are a number of different ways to produce light, with some more important in different parts of the electromagnetic spectrum than others. One universal factor that contributes to electromagnetic radiation over much of the spectrum is heat. All bodies produce electromagnetic radiation that, when we can feel it, we perceive as radiant heat. Examples of this include the electric

---

[1]The actual boundaries of the infrared are not well defined and depend to some extent on the technologies used to detect the radiation, as we will see in Chapter 2.

fire giving out a pleasantly warm red glow on a winter's day, or the tungsten lightbulb in my desk lamp shining down on my paperwork. The light we receive from the Sun is also radiant heat, shining across the 150 million kilometres of empty space that separate it from the Earth. All of these are obvious forms of radiant heat because we can detect it, both with our eyes and through the warmth it brings to our skin.

But everything else is putting out electromagnetic radiation, with a spectrum depending on its temperature, from the mirror surface of the Herschel Space Observatory at a temperature of -183 C to the entire universe, which is glowing with thermal radiation appropriate for its temperature of -270.27 C.

There are several ways of quantifying temperature, including both the Celsius and Fahrenheit scales. Physicists, though, prefer to use the Kelvin scale. This is similar to the Celsius scale used in much of the world, but instead of measuring temperature above the freezing point of water, at 0 C, it measures temperature above absolute zero. This is at a temperature of -273 C and is the lowest temperature anything can have, since it corresponds to the point at which there is no internal motion of all the atoms and molecules that make up a material, and no emission of thermal radiation. Nothing can ever reach a temperature of absolute zero, since there is always something warmer nearby radiating heat that will be absorbed by the colder body, heating it above -273 C. Absolute zero can, however, be approached for scientific purposes using heroic and ingenious cooling techniques. For the examples given above, the corresponding Kelvin temperatures are 90 K for the Herschel mirror and 2.73 K for the temperature of the universe. Room temperature, at around 20 C, corresponds to 293 K.

Thermal, or black body radiation as it is called, has a characteristic spectrum that arises because the electromagnetic radiation is in equilibrium with the matter it is in contact with. Over time, all systems will evolve to a state where equal amounts of radiation are absorbed and emitted at different frequencies, in equilibrium with their surroundings. A hotter body is more energetic, and so it emits more thermal radiation than a colder body.

The actual spectrum of black body radiation (see Figure 1.3), the power emitted at each wavelength, was a puzzle for physicists at the end of the 19th century. The light coming out of a fixed cavity was considered to arise from all possible wavelengths of light that could fit inside it. These can be thought of as a set of standing waves whose wavelengths are a fixed multiple of the size of the cavity, so they always have a value of zero at the walls. A standard result of the study of heat and energy, making up the theory of thermodynamics, states that each of these standing waves, or modes, should have the same energy. As you can tell from Figure 1.4, there are many many more high frequency modes, waves that go up and down many times from one side of the box to the other, than low frequency modes. These correspond to high frequency electromagnetic radiation, so they would represent ultraviolet light or beyond. This means that, under the assumptions of 19th century, classical physics, there should be a huge amount of ultraviolet light in all

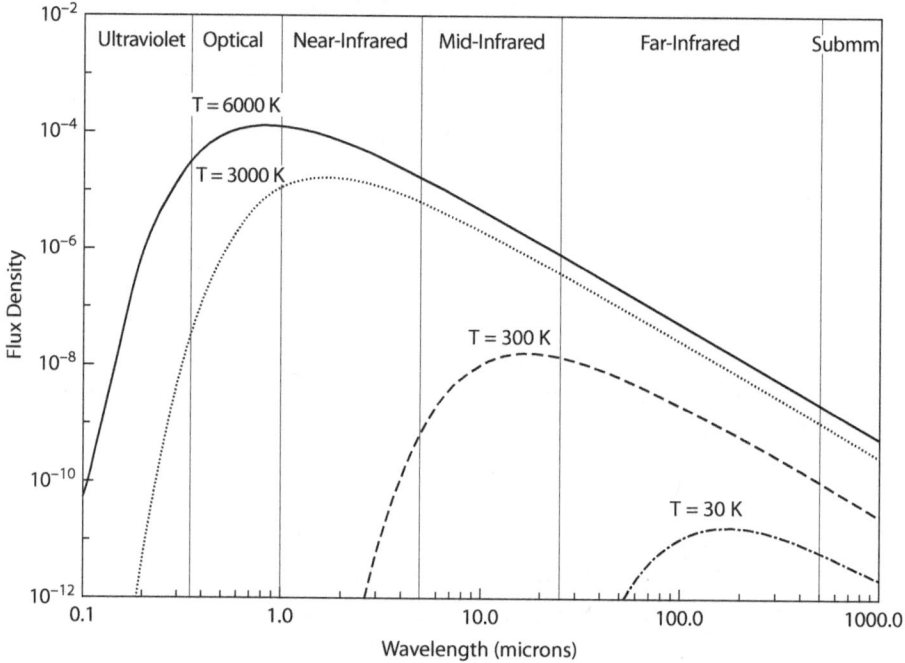

Figure 1.3 The spectrum of black body radiation at different temperatures.

Black body emission at different wavelengths for different temperatures. The black body spectrum for a star like our Sun, at a temperature of ∼6000 K, peaks in the optical. Cooler objects have black body spectra that peak at longer wavelengths. Something at room temperature, 300 K, emits most strongly in the mid-infrared, around 10 microns in wavelength, while interstellar dust, at around 30 K, emits most strongly in the far-infrared, around 100 microns.

thermal radiation. But we can easily see when we look at the spectrum of thermal radiation that this ultraviolet catastrophe, as it was known, is not the case.

The solution to this problem was devised by Max Planck in 1901, and provided the first hints of the quantum physics revolution that was to follow. Planck suggested that the amount of energy that could be carried by electromagnetic radiation was not continuous. Instead, this energy came in discrete lumps, or quanta, like tiny bullets.

Planck further suggested that the amount of energy each quantum of radiation could carry increased with increasing frequency. In effect, what he was saying was that electromagnetic waves have a particle-like as well as a wave-like nature, with the particles of light, which we now call photons, each carrying a fixed amount of energy given by their frequency multiplied by a constant, now given Planck's name - Planck's constant.

When this restriction on the amount of energy carried by photons of a given frequency is combined with the ideas of thermodynamics and the standing wave model for the production of thermal radiation, we get the black body spectrum seen in Figure 1.3 and measured across the world by 19th century physicists, even though they did not understand its origin.

The black body spectrum has several notable properties. Firstly, its peak wavelength depends directly on the temperature of the emitting body. The hotter it is, the shorter the wavelength of the peak, the colder it is, the longer the wavelength.

The surface of the Sun has a temperature of about 6000K, with a black body spectrum that peaks at wavelengths corresponding to the colour yellow. This, conveniently, thanks to evolution, is the wavelength at which the human eye has its peak sensitivity. Human skin, at a temperature of about 300K, emits black body radiation with a peak wavelength of about 10 microns, squarely in the mid-infrared, which is why mid-infrared cameras are used by emergency services to look for survivors in the aftermath of earthquakes and similar disasters. Most warm blooded animals also emit black body radiation peaking at a similar wavelength, as can be seen in Figure 1.5. Observations of the thermal spectrum of astrophysical bodies can thus tell us their temperature.

Secondly, since the amount of thermal energy radiated at a given temperature is constant no matter what the material is, we can also use the amount of energy received from a body to measure the amount of emitting material, given some other assumptions - though this doesn't work for bodies like planets that are opaque.

1.4   SPECTRAL LINES AND QUANTUM MECHANICS

Planck's solution of the ultraviolet catastrophe, by quantising the amount of energy in electromagnetic radiation, was just the first step towards a new way of looking at physics which solved many of the outstanding problems

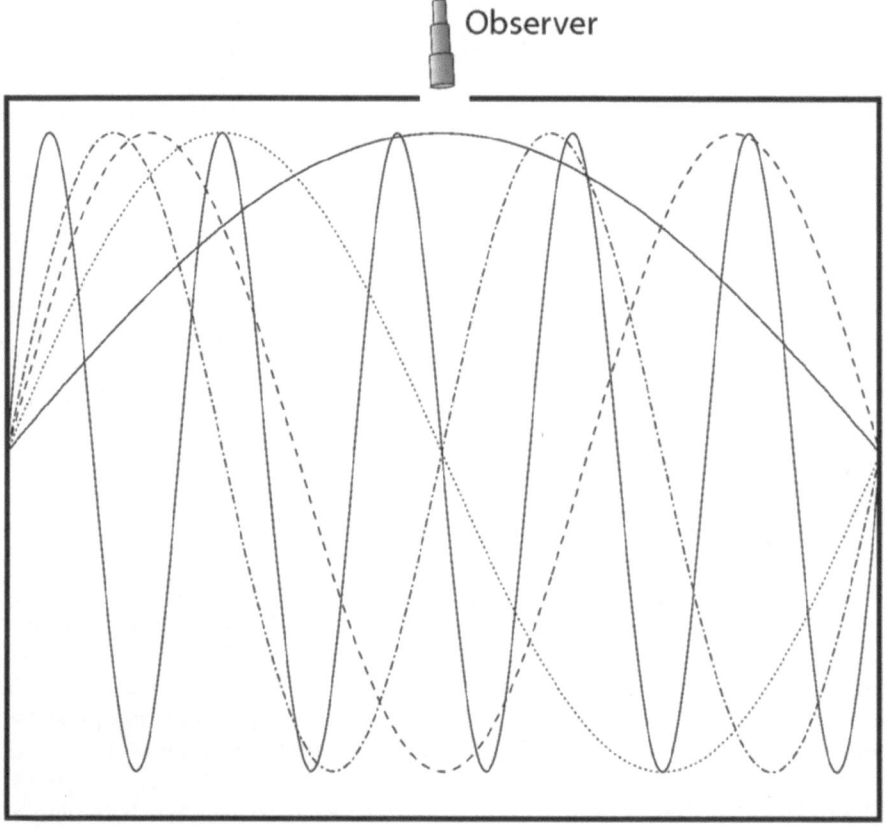

Figure 1.4 Standing waves in a cavity.
The standing waves that can fit inside a cavity of a given size, and which
would contribute to the black body radiation coming out of this cavity. There
are many more ways of fitting high frequency waves into a cavity of a given
size than there are of fitting in low frequency waves.

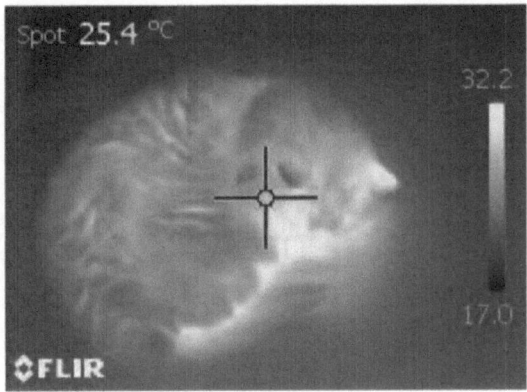

Figure 1.5 A cat, Domino, as seen in the mid-infrared.

in classical physics. Another of these, which is important in the context of astronomy, is the origin of spectral lines.

During the 19th century, scientists discovered that, in certain circumstances, a given material would emit light at specific wavelengths. This could come from a tube of gas through which an electric current was passed, or from a chemical compound being heated in the flame of a bunsen burner. The spectral lines seen in emission when a compound was heated could also be seen in absorption when light passed through a similar material. A tube of gas that gives out an emission line at a specific wavelength when excited by an electric current would, for example, absorb light passing through the tube at that same wavelength.

These spectral lines provided reliable fingerprints for identifying the presence of a given element, allowing chemical flame tests to determine what an unknown compound might contain. Copper, for example, produces a lovely green coloured flame in a bunsen burner. These techniques were applied to observations of stars and nebulae by William and Margaret Huggins (e.g. [80]). They found that astronomical objects contained the same chemical elements as we find on Earth.

The origin of spectral lines remained a mystery until the quantum physics revolution of the early 20th century. Planck's discovery that light is quantised, that it comes only in discrete particles whose energy depends on their wavelength, was only the start. Physicists soon realised that everything in the world of atoms and molecules is quantised. An atom or molecule can be in only a limited number of states. If an atom is hit by a photon of light whose wavelength matches the energy of one of those states, then the photon can be absorbed. Photons of other energies cannot be absorbed so easily. This process gives rise to spectral absorption lines at specific wavelengths, while the

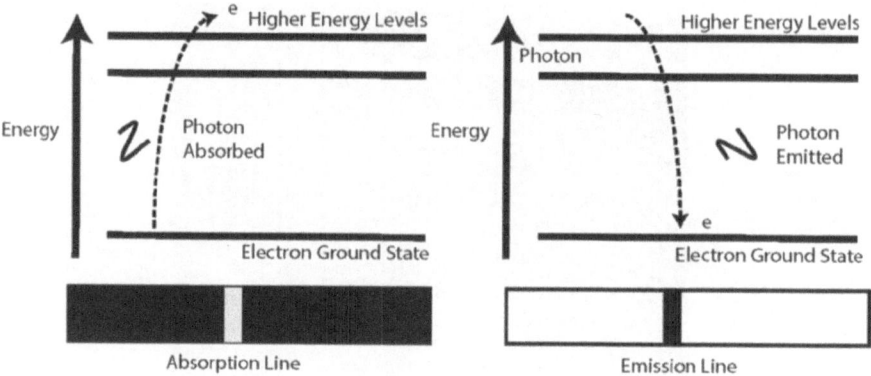

Figure 1.6 How energy levels produce spectral emission and absorption lines.

On the left, an electron in an atom absorbs a photon with energy appropriate to shift it from its lowest energy level, its ground state, to a higher energy, excited, state. This produces an absorption line. On the right, an electron in an excited energy state spontaneously decays back to its ground state by emitting a photon of the same, appropriate energy. This produces an emission line. The diagrammatic spectra shown below the energy level diagrams are negatives, in the sense that black means light is detected, and white means nothing is seen.

opposite process, whereby an atom decays from a high energy, excited state to a lower energy state, produces an emission line (see Figure 1.6).

The exact wavelengths of the lines depend on the detailed properties of the atoms, since the process of absorption and emission are shifting the electrons within the atoms from one allowed quantum state, also known as an energy level, to another. The atom doesn't pass through any intermediate states in such a transition - it just goes from one state to another in what has been called a quantum shift[2]. It isn't just atoms that have quantised transitions giving rise to spectral absorption and emission lines. The same physics applies to oscillations in molecules, where some of the atoms in the molecule bounce around like weights on a spring. Many of these molecular transitions are at lower energies, in the infrared rather than the optical, since the absorption of higher energy photons may break the weaker bonds that keep molecules in one piece.

Observations of spectral lines, both emission and absorption, can be used

---

[2]This transition from one state to another, different, state through a quantum shift has led to this term 'quantum shift' being used by politicians and others to mean a fundamental change. This has caused some hilarity among physicists since, at some level, a quantum shift is in fact the smallest change it is possible to make. Of course, it may be that politicians know this is the case, and are hoping the public will not notice.

in astronomy for more than just working out what elements and molecules make up an astronomical object. Emission lines from atoms in higher energy, excited states can reveal something about the nature of the radiation that is exciting them. Ionised material, atoms and molecules that have had electrons stripped away from them, have different spectra than the un-ionised versions. High levels of ionisation, where an atom has had many of its electrons stripped away, take a lot of energy, possibly implying the presence of hard, X-ray radiation or some other highly energetic processes which might not be directly detectable by an observer. The study of spectral lines, across the electromagnetic spectrum, is thus a very powerful tool for the observational astronomer.

1.5 THE STUFF OF THE UNIVERSE

Having looked at the electromagnetic spectrum, and how light is produced or absorbed, it is time to look at the other constituents of the universe - the stuff that we are made of and which drives the activities of stars, planets and galaxies - since it is their behaviour we are trying to determine through astronomical observations, in the infrared and at other wavelengths.

We have already mentioned atoms and molecules. Atoms are the building blocks of chemistry. Each atom is made of a different element, and these range in mass from hydrogen, the lightest element, to uranium, the heaviest naturally occurring element, and beyond, to the small number of heavier elements created artificially in nuclear reactors and particle accelerators.

The most common elements in the universe are hydrogen and helium. These elements emerged from the Big Bang at the start of the universe, along with a small amount of slightly heavier elements such as lithium and beryllium. All the other chemical elements, including the carbon, oxygen, nitrogen, calcium, etc. that we find in our bodies, were made inside stars over the 13.8 billion year age of the universe.

Molecules are atoms combined together by chemical reactions. They can be very simple, such as molecular hydrogen, which consists of two hydrogen atoms linked together, or very complex, such as the DNA molecules in the cells of our bodies that carry genetic information. The types of molecules seen in astronomical objects are usually quite simple, collections of just a few atoms, since there are few opportunities for large numbers of atoms to gather together to form complex molecules in the vacuum of space, and because ultraviolet light from stars has a tendency to break complex molecules apart. The most common molecules in space are molecular hydrogen, two hydrogen atoms joined to each other, and carbon monoxide.

We can detect atoms and molecules in astronomical objects through their spectroscopic absorption and emission lines. These lines occur at many different wavelengths, corresponding to the energies needed to excite them in various ways. For atoms, this would mean moving electrons up and down their different energy levels. For molecules, various different vibrations between their constituent atoms lead to similar quantised energy states. Molecular transitions are generally quite low in energy, so their corresponding spectral lines lie in the near- mid- and far-infrared.

Atoms themselves are made up of two components: nuclei and electrons. Nuclei are small but contain almost all of the mass of an atom. They are about 10^{-15}m across, compared to the size of an atom, which is about 10^{-8}m across - nuclei, in other words, are less than a millionth the size of an atom. Electrons form a cloud around the nucleus, defining the size of the atom, and controlling its chemical reactions. The electrons are bound to the nucleus thanks to their electric charges, with electrons having a negative charge, and the nucleus a positive charge, meaning there is an attractive electrostatic force between them.

So far, electrons appear to be fundamental particles, incapable of division into smaller components. This is not the case for atomic nuclei, which are made up of two kinds of particles, protons and neutrons, whose mass is roughly equal. Protons have a positive electric charge, equal to that of the electron, while neutrons have no charge at all.

The number of protons in a nucleus determines which chemical element the atom represents, since it also determines the number of electrons needed to surround the atom to make it electrically neutral. There should be an equal number of electrons and protons or the atoms will have an electric charge. Electrons can be knocked off atoms in a number of ways, through collisions with other atoms, for example, or through the absorption of a photon with high enough energy. If this happens we call the atom an ion, and can say that it has become ionised (see Figure 1.7).

Not all combinations of protons and neutrons in a nucleus are stable. In fact, most combinations are likely to be highly unstable and will rapidly decay into something else with a burst of radiation. Where a mixture of protons and neutrons is stable enough to exist for an appreciable amount of time, those species with the same number of protons but different numbers of neutrons are called isotopes.

Hydrogen is the simplest element, and usually comes with just one proton as its nucleus, orbited by a single electron. A second stable form of hydrogen also exists, with a neutron added to the proton in the nucleus (see Figure 1.7). This isotope, known as deuterium, would still be orbited by a single electron, meaning it would have the same chemical properties as hydrogen, but its nucleus, with a proton and a neutron, is twice as heavy. Deuterium, just like normal hydrogen, can react with oxygen to form water, H_2O, but water made with a deuterium atom, sometimes given the symbol HDO, has a heavier nucleus, and thus is often called heavy water. There is also a third kind of hydrogen, where the nucleus has two neutrons and a proton, called tritium. However, the tritium nucleus is unstable, and decays through the conversion of one of its neutrons into a proton and the emission of an electron and a neutrino (a chargeless, nearly massless particle). This leaves behind a nucleus with two protons and one neutron, forming the nucleus of an atom of helium, the next lightest element to hydrogen. The decay of any given tritium atom

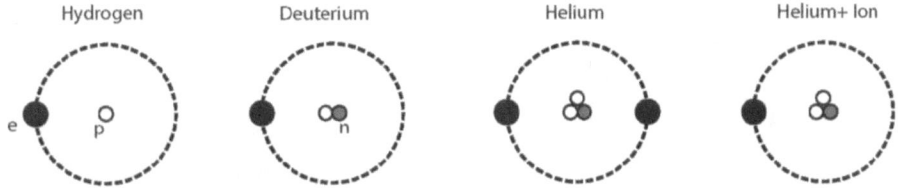

Figure 1.7 Diagrammatic representation of atoms of hydrogen, deuterium and helium, and an ion of helium.

On the left is a simple hydrogen atom, consisting of a single proton (p) as its nucleus, orbited by a single electron (e). Next is an atom of deuterium, an isotope of hydrogen, which has a proton and a neutron (n) in its nucleus. Then we have the next heaviest element, helium, which has two protons in its nucleus and a single neutron, orbited by two electrons. There is also an isotope of helium with two protons and two neutrons in its nucleus. Finally, on the right, we have a helium atom that has lost an electron, and thus has become an ion with a single positive charge. If it were to lose its other electron then it would be an ion with two positive charges. Note: The nucleus and electrons are not shown to scale. The nucleus is in fact over a million times smaller than the size of an atom, which would make it difficult to see if drawn to scale.

happens at random, but the rate of tritium atom decays is such that half the tritium atoms you might have will have decayed to helium after a period of 12.3 years - this is the half-life of tritium[3].

Atoms transform from one element to another through radioactive decay, but they can also be transformed into other elements through two other processes, fission and fusion (see Figure 1.8). Fission is the process that takes a large atomic nucleus, like that of uranium, and splits it apart. This can be triggered when the nucleus is hit by a neutron of sufficiently high energy. The result can be quite a mess, with nuclear fragments of different sizes, and the release of a lot of energy and further high energy neutrons. These neutrons can, in principle, go on to trigger further fission reactions in other uranium nuclei, creating a chain reaction. This is the process that powered the first atomic bombs.

Heavy elements like uranium are exceedingly rare in the universe, so fission reactions play no significant role in astrophysics. Fusion reactions, in contrast, play an incredibly important role. Fusion is the opposite of fission. Instead of breaking large nuclei apart, fusion takes small nuclei and releases energy by

---

[3]While the chemical reactions of deuterium and tritium are the same as normal hydrogen, the different mass of the resulting atoms slightly changes the properties of chemical compounds that include deuterium rather than hydrogen. While not actively poisonous, a diet of heavy water, rather than normal water, would not be good for you in the long term. Since tritium is radioactive, decaying with a half life of 12.3 years, a diet of tritium instead of normal water would be highly dangerous!

Hydrogen Fusion to Helium   Nuclear Fission

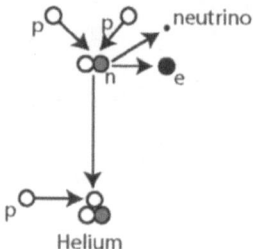

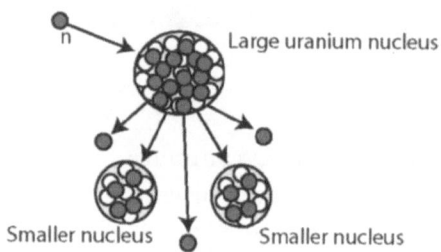

Figure 1.8 Diagrammatic representation of nuclear fusion and fission. On the left is a diagram of the fusion process by which most stars get their energy. A proton - a hydrogen nucleus - hits another proton and they convert into a deuterium nucleus through the emission of an electron and a neutrino. Then, another proton hits this deuterium nucleus, which transforms into a helium nucleus, along with the emission of energy in the form of gamma ray photons. In fission, a large nucleus, like that of uranium, is hit by a neutron, triggering the splitting of the uranium nucleus into two smaller nuclei, with the emission of both energy and further neutrons. These neutrons may go on to trigger further fission reactions in other uranium nuclei.

getting them to stick together. You can, for example, take hydrogen nuclei and smash them together to form helium. This fusion reaction takes place in the heart of stars and provides much of the energy in the universe.

How do all these transformations of atomic nuclei work? How, for example, can a neutron in a tritium atom convert itself into a proton, an electron and a neutrino?

All of these reactions inside atomic nuclei, and much more besides, are possible because protons and neutrons are not fundamental particles. Instead, they are composite entities, made up of collections of smaller entities called quarks. There are six kinds of quarks, given the names up, down, strange, charm, top and bottom[4]. The up and down quarks, though, are responsible for almost all of the physics that we see outside particle accelerators and the early universe. Protons are made of two up quarks and one down quark, while neutrons are made of two down quarks and one up quark (see Figure 1.9). The process that turns a neutron into a proton, with the emission of an electron and a neutrino, is driven by the transformation of a down quark into an up quark.

There is one other constituent of of the universe that needs to be discussed,

---

[4]These last two have the alternate names truth and beauty, which, to be honest I find rather more attractive.

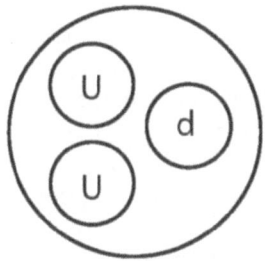
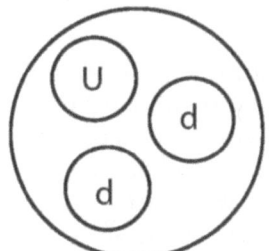

The quarks inside a proton   The quarks inside a neutron

Figure 1.9 The internal structure of protons and neutrons.
Protons are made up of two up quarks and a down quark, while neutrons are made up of two down quarks and an up quark.

and that is dark matter. This is difficult to explain because we don't *yet* know what it is.

However, we do know what dark matter *isn't*. As will be explained in Chapter 8, we know it isn't made up of normal matter, built of protons, neutrons and other particles that consist of three quarks and that are generically called baryons.

We also know what dark matter does. Like normal, baryonic matter, it has mass, and attracts things gravitationally. It doesn't do this on scales where we can easily see the effects - it won't, for example, affect apples falling out of trees - but the gravity of dark matter has a strong influence on the motions of stars in galaxies, and galaxies within galaxy clusters and larger scale structures. Determining the nature of dark matter, and how it might relate to fundamental theories of physics like string theory, is one of the major goals of both cosmology and particle physics at the moment.

There is also another player in the way the universe works on the largest scales, *dark energy*. This is even more mysterious than dark matter. We can only see its effects on the largest scales of both distance and time, where it acts to increase the rate at which the universe is expanding. The physics behind dark energy is very uncertain, and its discovery came as a great surprise in the late 1990s.

1.6 A VERY BRIEF HISTORY OF THE UNIVERSE

We will learn a lot more about dark energy and dark matter when we look at cosmology in Chapter 8, but before we move on to more practical matters of astronomy we need to know a little bit about the kind of things that astronomers observe, whether in the infrared or at other wavelengths.

The universe began with the Big Bang, an explosion not only of matter but of the space-time in which matter exists. The universe is still expanding, with the distances between galaxies increasing all the time. If there were no dark energy, the expansion rate of the universe would be slowing as a result of the gravitational attraction of all the matter inside it. However, we can see that the expansion rate is not slowing, it is in fact increasing, and this effect is down to dark energy. This will be discussed more in later chapters, mainly Chapters 7 and 8.

The distribution of matter in the universe was initially highly uniform, but there were small perturbations, with some regions slightly denser than others, at a level of about one part in a hundred thousand, produced by quantum fluctuations at the earliest stages of the universe. Those regions that are slightly denser slowly grew as their gravity attracted the matter around them. At this early stage of the universe nearly all matter was in the form of atomic hydrogen.

The densest perturbations eventually attracted enough matter to begin a runaway collapse leading to the formation of the first stars, powered by the fusion of hydrogen into helium. These first stars are thought to have been much more massive than the stars we see today. They burned very brightly, ionising the atomic hydrogen around them, and rapidly ending their lives in giant supernova explosions that spread material enriched with elements heavier than hydrogen and helium into the space around them.

While the first stars were forming, the dense regions they inhabited would have continued to grow, attracting more hydrogen gas from the surrounding universe. The enriched material from the first stars made it easier for this accreted material to cool, leading to the first generation of stars similar to those we see in our own galaxy today.

As further generations of stars formed and died, they recycled ever more enriched material into their parent galaxies. Eventually, there were enough heavy elements, like carbon, silicon, oxygen, nitrogen and iron, that dust grains could form in the space between the stars. This dust was swept up in future generations of star formation, leading to the formation of rocky bodies circling new generations of stars. These rocky bodies are planets.

The Earth, our atmosphere, and all the atoms in our own bodies can be traced back through these cycles of star formation, all the way to the Big Bang.

1.7 THE PRACTICAL: TELESCOPES

Astronomy is an observational science, so we need tools to observe the sky. The next chapter looks at how light, and in particular infrared light, is detected, but before we can make useful detections that light has to be collected. For that we need telescopes.

There are two main types of telescope: those that collect light using lenses, called refracting telescopes, and those that collect light using mirrors, called reflecting telescopes. The first generation of telescopes, used by pioneering astronomers like Galileo, were refracting telescopes, but nearly all professional

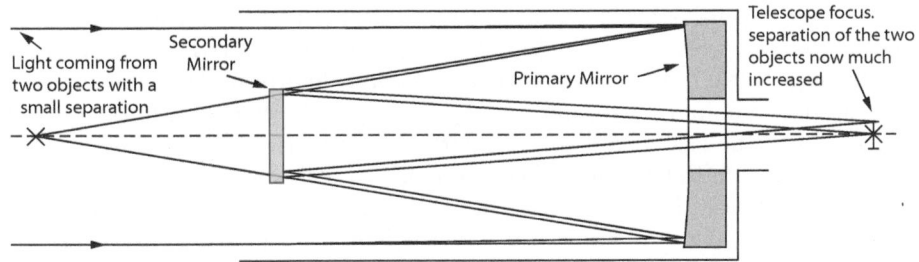

Figure 1.10 Diagram of a typical reflecting telescope.
The large primary mirror of a telescope not only collects light over a large area, increasing the sensitivity, but also magnifies the angular separation of objects that would otherwise be indistinguishable. (Modified from an original public domain image by Griffenjbs.)

telescopes today, and many of the best amateur telescopes, are reflectors. There are several reasons for this, but one of the main ones is the need to collect more light to see fainter objects. While both lenses and mirrors get more difficult to handle the larger they get, using lenses becomes more problematic more quickly since they can only be supported around their outside. This means they tend to bend and stretch under their own weight. Once that happens, they are no longer able to keep astronomical objects properly focused.

This is not to suggest that lens-based telescopes or binoculars are useless for astronomy - quite to the contrary! Galileo revolutionised our view of the universe with a refracting telescope, and the easiest introduction to astronomy is probably to use binoculars. Professional astronomers, though, need something more precise and with a far larger collecting area. Glass lenses also have a problem in the context of infrared astronomy since glass is opaque at many infrared wavelengths.

The layout of a typical reflecting telescope is shown in Figure 1.10. This shows both how the large, primary mirror on the right collects light over a large area, and how the focussing of light enlarges the angular differences between light rays coming from two close but different objects. The larger the mirror, the smaller the angular separations that can be measured. There is another factor here, however, in that longer wavelength light leads to inherently less distinct images. To see the same angular scales in the longer wavelength near-infrared as in the optical, therefore, you would need a mirror perhaps twice as big. Sensitivity also leads to the need for larger mirrors, since they collect more light, making fainter objects more detectable.

Larger mirrors, then, would seem to be essential, but practical considerations set some limits on our ambitions. Space based telescopes have to be launched into orbit. This means that the mirrors have to fit inside the launch vehicle, setting a limit on the size of the primary mirror (though we will see in Chapter 9 plans for how one telescope will get around this limitation). Ground-based telescopes have limitations of their own. Firstly, the mirror has to be kept in the right shape so that it doesn't deform under its own weight as the telescope moves around the sky. For optical telescopes, until recently, this limited the size of a single primary mirror to less than about 5 metres across. Over the last twenty years, though, two technologies have come along that have allowed the construction of optical and near-infrared telescopes with mirrors up to 10m in diameter. The methods used include the active support of a large, thin mirror, and the construction of a large mirror out of smaller hexagonal segments. We will see examples of both kinds of telescopes in the next section. Telescopes dedicated to longer wavelengths do not need the same precise mirror surface as an optical telescope. Much larger mirrors in the radio and in the far-infrared/submillimetre are therefore possible.

The other thing that an astronomical telescope needs is a clear sky, free of clouds or other obscuration, and free of light pollution that might otherwise outshine the stars themselves. This is clearly not a problem if your observatory is in space, but on the ground it means that astronomers seek out the tops of mountains in isolated places, far away from the rest of civilisation. The fact that one of the best such sites is on one of the Hawaiian islands is merely a happy coincidence.

1.8   OBSERVATIONAL STARS

Having looked at some of the physical and engineering constraints on building telescopes, it is time to look at some of the major players in the world of observational infrared astronomy. Some of these telescopes are on the ground, and others are in space. Many of the space-based observatories have reached the end of their operational lives and no longer function, but their influence remains. Many of the most influential of these telescopes will be discussed in more detail in later chapters, but here is a *dramatis personae* of some of the telescopes that have made infrared astronomy what it is today.

1.8.1   Ground-based telescopes

The United Kingdom Infrared Telescope (UKIRT) (Figure 1.11) was one of the first large telescopes (3.9 m diameter primary mirror) to be dedicated to, and optimised for, observing in the infrared. Built on the summit of Mauna Kea in Hawaii, it began operation in 1979 and has been at the forefront of near- and mid-infrared astronomy ever since. Most recently, UKIRT has been dedicated to a series of sensitive near-infrared surveys over large areas of the sky. We will see the results of some of these surveys later in this book. A similar, slightly smaller (3 m diameter mirror), infrared telescope called the Infrared Telescope Facility (IRTF) was built by NASA on Mauna Kea at the same time as UKIRT. IRTF was originally built to support the Voyager missions to the outer planets of the Solar System, but has since become a general purpose

Figure 1.11 UKIRT: the UK Infrared Telescope.
(Courtesy UKIRT/JAC, from http://www.jach.hawaii.edu/.)

Figure 1.12 The Keck Telescopes.
Two of the largest optical/infrared telescopes in the world.

infrared telescope, albeit with capabilities that make it especially well suited to observations of planets and other Solar System objects.

The Keck telescopes (Figure 1.12) are also built on Mauna Kea, and are among the largest optical and near-infrared telescopes in the world, with primary mirrors that are 10 m in diameter. This large mirror size is made possible by using segmented mirrors, made up of 36 hexagonal mirror segments. The first Keck telescope was completed in 1990 and started full scale science operations in 1993. It was the first telescope in the world to use a segmented mirror. This was considered a risky undertaking at the time, but the design was such a success that a further donation from the WM Keck foundation allowed a second giant telescope to be built, Keck 2, which began operation in 1996. While the Keck telescopes are not dedicated to infrared astronomy, observations in the near-infrared take up a substantial fraction of the time available on the two telescopes, and about half of the instruments at the observatory operate in the infrared.

The European Southern Observatory (ESO) Very Large Telescope (VLT, Figure 1.13) lies atop Cerro Pachon in the Atacama desert in Chile. In some sense the name 'Very Large Telescope' is something of a misnomer since the VLT is actually made up of four separate 'unit telescopes', each of which has

Figure 1.13 The European Southern Observatory Very Large Telescope. The four 8.2m unit telescopes of the VLT, together with the four smaller auxiliary telescopes. (Courtesy of ESO; http://www.eso.org/public/.)

an 8.2 m diameter actively supported thin mirror. Much of the time the VLT operates as four separate telescopes, but it is also possible to combine the light from the four unit telescopes, together with light from four smaller, 1.8 m auxiliary telescopes. The light from these different telescopes is combined in such a way that the light rays interfere with each other to provide angular resolutions that are comparable to what would be possible with a single mirror with a size equal to the separation between the individual telescopes. In the case of the VLT this distance is about 100 m. This system is known as an interferometer, and it can observe much smaller structures than the individual 8 m VLT telescopes. Interferometry is actually a much more usual technique at longer submm and radio wavelengths, and the VLT interferometer, combining light from so many individual telescopes, is one of its most innovative features.

Like the Kecks, the VLT functions as both an optical and an infrared telescope, but many of its instruments operate in the near- and/or mid-infrared. The adoption of infrared instrumentation by observatories that might once have been thought of as traditional optical observatories is an example of how infrared astronomy has become part of mainstream astronomy over the last twenty-five years. In the past, infrared astronomers had to go to UKIRT or IRTF for their data. Now they can go to pretty much any observatory. These include other 8 m-class telescopes, like Subaru or the Gemini telescopes, the Gran Telescopio Canarias, which, at 10.4 m, is currently the largest single aperture optical/infrared telescope in the world, or the smaller telescopes at observatories like ESO's La Silla Observatory.

Observers operating at longer wavelengths in the submillimetre still have to go to dedicated observatories. Among these is the James Clerk Maxwell Telescope (JCMT, Figure 1.14), which is another resident of Mauna Kea. Since it is operating at much longer wavelengths than optical/near-infrared telescopes like UKIRT or VLT, it can get away with a much poorer surface accuracy and thus can have a 15 m diameter mirror, much larger than can

Figure 1.14 The James Clerk Maxwell Telescope.
The 15 m James Clerk Maxwell Telescope in Hawaii. (Courtesy of JCMT/JAC, http://www.jach.hawaii.edu/.)

currently be built in the optical. This mirror can be seen in Figure 1.14, and you can tell that it is not the kind of bright shiny mirror that you might be used to looking at in the bathroom. Close up, the mirror of the JCMT actually looks more like well-kept metal than a mirror. However, the JCMT usually doesn't appear as it does in Figure 1.14 since the telescope is usually hidden behind a large shroud of fabric, made from GoreTex. This acts as a wind shield, allowing the telescope to operate in higher wind conditions than would otherwise be possible. GoreTex is transparent at submm wavelengths but opaque in the optical, so the wind shield also acts to filter out daylight, allowing the JCMT to observe during the day as well as at night.

The JCMT is not the only large single-dish telescope that operates at millimetre and/or submillimetre wavelengths. Among others, there is the Caltech Submillimeter Observatory (CSO), which sits next door to JCMT on Mauna Kea. Operating at slightly longer wavelengths are the IRAM (Institut de Radioastronomie Millimetrique) 30 m telescope based in Spain at Pico Valetta, and the Large Millimetre Telescope (LMT) which operates on Sierra Negra, the fifth highest mountain in Mexico. The LMT has a 50 m diameter primary mirror, making it the largest single dish millimetre telescope in the world.

Combining the signals from different telescopes to form an interferometer is much easier at submillimetre and millimetre wavelengths than in the optical, so there are also several observatories based on this approach. The Submillimeter Array (SMA, see Figure 1.15) sits with the JCMT and CSO in what is called Millimetre Valley on Mauna Kea. The SMA combines the signals received from eight separate six-metre diameter telescopes to produce images with a resolution comparable to what would be achieved by a dish up to 500 m across.

Other interferometers operating at millimetre and submillimetre wavelengths include the Plateau de Bure interferometer in the French Alps operated by IRAM, and the Combined Array for Research in Millimetre-wave Astronomy (CARMA) in the Inyo mountains of California. The biggest and best such interferometer, though, is the Atacama Large Millimetre Array (ALMA), currently nearing completion at an altitude of 5000 m in the Atacama desert of Chile. We will learn more about ALMA in Chapters 7 and 9.

1.8.2 Space-based observatories

Observations from the ground, as we shall see in the next chapter, are impossible at many infrared wavelengths, and very difficult at others. To escape these difficulties, telescopes have to be sent into space. This is difficult and expensive, but the scientific results from these space-based observatories have often been stunning and have produced revolutions in our understanding of the universe.

The first space-based infrared observatory was the Infrared Astronomy Satellite (IRAS, Figure 1.16)) and a lot more will be said about IRAS in Chapter 6. In many ways IRAS was the foundation of all far-infrared astron-

Figure 1.15 Millimetre Valley on Mauna Kea.
The three millimetre/submillimetre telescopes of Millimetre Valley. From left to right the CSO, the JCMT and both the control building and the separate dishes of the SMA.

omy that has come after it. One of the key products of the IRAS satellite was a catalog of infrared sources at wavelengths of 12, 25, 60 and 100 microns over the entire sky. IRAS was launched in 1983, but this catalog is still used almost every day by infrared astronomers. This is quite an achievement when you realise that IRAS's primary mirror was only 57 cm across and the satellite only operated for ten months before its liquid helium coolant ran out and its detectors stopped working.

After IRAS, the Hubble Space Telescope (HST) was the next infrared-capable observatory to be launched. While primarily an optical observatory, HST has been equipped with several different near-infrared instruments over the course of its lifetime. These include the NICMOS (Near Infrared Camera and MultiObject Spectrograph) instrument, which operated from 1997 to 1999 and, following an upgrade to its cooling system, from 2002 to 2008. A second infrared-capable instrument was installed in 2009, the Wide Field Camera 3 (WFC3, see Figure 1.17). The HST has a main mirror that is 2.5 m across, so was able to produce many sensitive and highly detailed images with both NICMOS and WFC3, some of which will be discussed later in this book.

Following the success of IRAS there was a series of further space infrared observatories including the Infrared Space Observatory (ISO), Spitzer (see Figure 1.18) and Akari, launched by the European Space Agency (ESA), NASA and Japan, respectively. ISO and Akari were direct successors to IRAS in many ways, and all three of these observatories had relatively small primary mirrors (Spitzer being the largest at 85 cm) cooled by liquid helium. ISO was the first

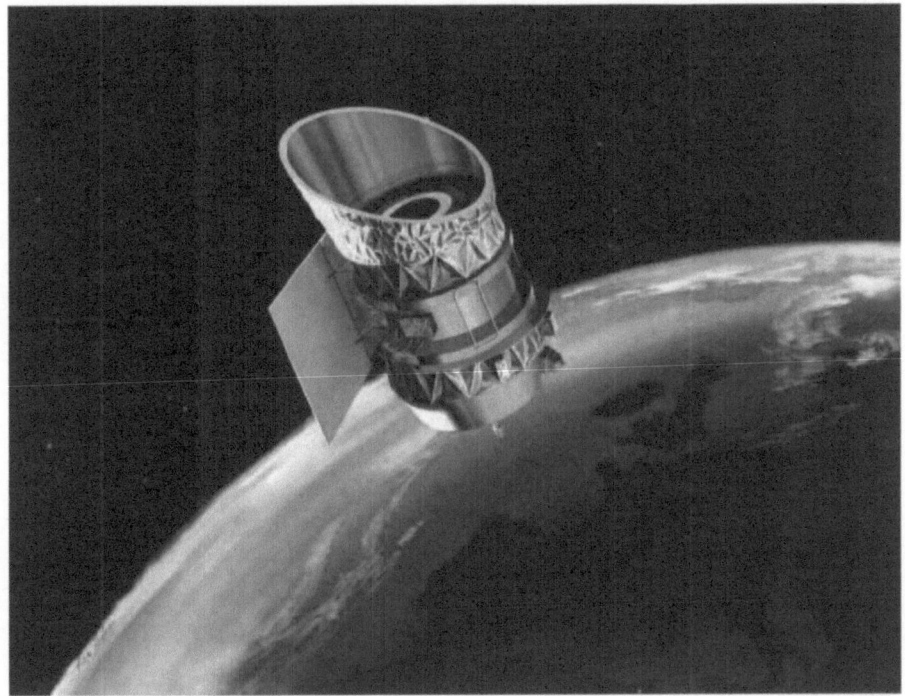

Figure 1.16 The IRAS Satellite.
An artist's impression of the IRAS satellite while operating in Earth orbit.
(Courtesy of NASA.)

Figure 1.17 The WFC3 Instrument.
The Wide Field Camera 3 about to be installed by astronauts on the Hubble
Space Telescope. (Courtesy of NASA.)

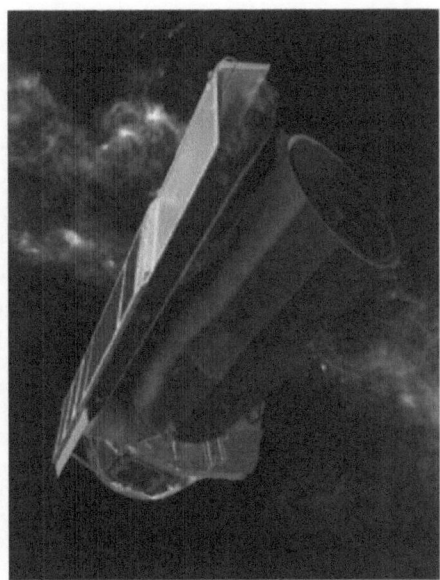

Figure 1.18 The ISO and Spitzer Satellites.
(Courtesy of ESA and NASA.)

to be launched, in 1995, with Spitzer following in 2003 and Akari in 2006. ISO's supply of helium ran out after about two and a half years, while Akari's coolant lasted about 18 months, though its instruments in the near-infrared continued operation until May 2011 when it suffered a serious electrical failure and ceased operations. The coolant aboard Spitzer lasted rather longer as it was launched into a more distant orbit than the other satellites. This meant it was not constantly bathed in the warm glow of reflected sunlight from the Earth and could continue full operation with its liquid helium coolant for nearly 5 years 10 months. The coolant ran out in 2009, and since then Spitzer has continued to operate in 'warm' mode, taking images at wavelengths of 3.6 and 4.5 microns.

The most recently launched infrared satellite mission was the WISE satellite (Wide-field Infrared Survey Explorer). This was launched in December 2009 and conducted a survey of the entire sky in the near- and mid-infrared, at wavelengths of 3.4, 4.6, 12 and 22 microns. This survey ended early in 2011, but the satellite has recently been reactivated by NASA to search for potentially hazardous asteroids.

The history of space-based infrared astronomy ends with two of the most ambitious missions to date, the Herschel Space Observatory and the Planck spacecraft . These were launched together in May 2009. At 3.5 m, Herschel has the largest telescope mirror ever launched into space, while Planck's 1.5 m mirror was used to survey the entire sky to study the light left behind by the Big Bang. Much more will be said about these projects in Chapters 7 and 8.

1.9 CONCLUSIONS

William Herschel's discovery of infrared light in 1800 was the starting point for many things. It began the series of investigations of electromagnetic radiation that would lead, over the course of the 19th century, to the revolution in our understanding of the world that is quantum mechanics. This is probably not something that Herschel expected since he was far more of an astronomer than a physicist. More slowly, it led to the development of astronomy at wavelengths other than the optical.

The reason Herschel's discovery of infrared light didn't immediately lead to infrared astronomy is because there was no easy way to detect such light until much more recently. Herschel's approach of using thermometers only worked with sources as bright as the Sun, although the physics behind that approach, the measurement of temperature increase as a result of absorbed radiation, is behind some of today's most advanced and sensitive detector technology.

The next chapter looks at how light, and especially infrared light, is detected, and at some of the practical problems that send infrared astronomers to the tops of mountains, and their telescopes into orbit and beyond.

CHAPTER 2

Feeling the Heat

2.1 THE PASTY DETECTOR

It looked like a small, gold painted waste paper bin, with pipes coming out the top, and a big fat lens sticking out of one side. It was the instrument that was meant to be taking the data for my PhD, and, perhaps a year late, it was just beginning to work.

Inside the cylinder were a series of nested enclosures. The one nearest the top held liquid nitrogen. This was filled by a tube that emerged to one side of the metal plate that sealed the top of the instrument. In the middle was a cylinder of liquid helium, connected to the tube at the centre of the top plate. This was closed off with an elaborate mixture of rubber and metal pipes, designed to allow boiling helium out, but to prevent air falling into the tube and freezing solid. Below the helium reservoir, sealed inside a radiation shield to make sure that nothing was exposed to anything above liquid helium temperatures, were the optics, detectors and the little mini-fridge that had just finished cooling them to a fraction of a degree above absolute zero.

Simon, the instrument's designer, connected the readout electronics, looking pleased that we now had something that might just work.

'Let's see if it works', he said, reaching over for one of the hot pasties that a fellow graduate student had just given us for lunch.

He placed a volt meter on the bench in front of us, and waved the pasty in front of the opaque, plastic lens.

The needle rose, going from one side of the dial almost all the way over. Simon nodded with pride.

'See', he said. 'We've made a pasty detector'.

2.2 THE HISTORY OF ASTRONOMY: THE HISTORY OF DETECTION

Science is about measurement. If you can't measure something, then it is very difficult to do any science with it. The history of infrared astronomy, and of

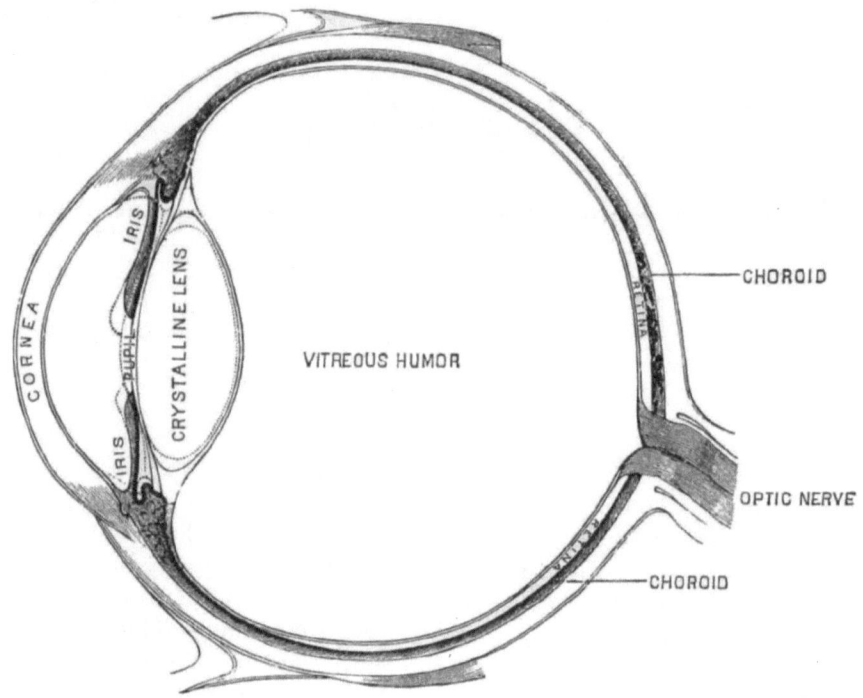

Figure 2.1   The Mark One Eyeball.
An anatomical diagram of the human eye, showing the lens and retina [59].
(From: *Popular Science Monthly*, 45, 1894.)

astronomy more generally, has been determined by what kinds of signals we can detect from the rest of the universe. In this chapter we will be looking at how infrared light is detected.

Astronomy is almost unique among the sciences in that just about everyone is equipped from birth to be able to make astronomical observations using what is technically known as the Mark 1 Eyeball. Indeed, for most of the history of astronomy, from the builders of Stonehenge, to the Babylonians and ancient Greeks, right up to the 19th century, the human eye has been the mainstay of astronomical observation. The eye, in fact, shares many of the general features of other astronomical instruments. Understanding various aspects of its operation will be useful in understanding the artificial and more specialised detectors we will be looking at later in this chapter. It will also reveal the limitations that mean that the human eye is not that useful for modern astronomy even in the optical part of the electromagnetic spectrum where it operates.

Figure 2.1 shows an anatomical drawing of the human eye, with the various parts, from pupil to retina, labelled. The eye works by collecting light entering through the pupil, and then focusing this light, through a combination of the lens and the aqueous humour at the front of the eye, to form an image on the retina. The bulk of the eye functions like a small telescope, or like the optics of the camera built into a mobile phone. The image formed by these optics is detected by the retina. This is where the photons of incoming light are converted to nerve signals that are then communicated to the brain via the optic nerve. Two different types of cells in the retina, known as rods and cones, are responsible for detecting light. The rods are capable of detecting fainter light levels, but have no colour sensitivity, while the cones require more light to function, but are divided into three types, in most people, providing sensitivity to the separate colours red, green and blue. The cones are predominantly arranged in a densely packed region of the eye called the fovea, allowing the high resolution sight required for precision examination of things at which we are directly looking. This is necessary, for example, when we're reading something, such as this book. The rods, in contrast, are distributed away from the fovea. This is why when you need to see something in low light levels, which is often the case for astronomers, it is helpful to look away from what you want to see, so that the higher sensitivity rods around the edges of the eye can be used instead of the lower sensitivity cones. A side effect of this is that it is difficult to perceive colour in low light conditions, and is why you need a decent telescope to be able to see that stars have a wide range of colours.

The detection of light by rods and cones follows a similar process. These cells all contain photoreceptor proteins. When these proteins absorb a photon of an appropriate wavelength they break down into two constituent products, triggering a series of chemical reactions that leads to a signal in the optic nerve. In the rods the photoreceptor protein is rhodopsin, while the cones use a class of protein called photopsins. Usually there are three different types of photopsin in the human eye, sensitive to red, green and blue, and allowing us to see colours. People who are colourblind may have fewer types of cones, and thus lack some aspect of colour vision, and there are reports that a small number of people may have more than three types of cones, allowing them to see in four, or more, colours [87].

There are a number of problems with the human eye as an astronomical instrument, despite the fact that it was the only astronomical instrument that existed for most of the history of astronomy. One of the key problems is that the detection process does not stop with the triggering of a neurone in response to a rod or cone receiving an incoming photon of light. Instead, there is a very complicated, and still not fully understood, series of processing steps in the visual cortex of the brain before we get to perceive what the eye has seen [73]. These complex layers of processing have been optimised by evolution to allow humans to operate in their natural environment, and the needs of that environment are very different to the needs of a scientist trying to make an objective astronomical observation. This means, for example, that reaching quick decisions about potential hazards is essential, and that false alarms are better to have than missed hazards. It's a lot better if you jump away from what you think is a tiger than if you ignore a real tiger and get eaten. An example of such a 'false positive' in astronomy is the 'discovery' of the canals of Mars, where observers like Schiaparelli and Burton mistakenly saw linear features on the surface of the nearest planet. This resulted, at least in part, from the tendency of the human eye to see lines joining the kind of indistinct point features, such as craters, that are found on the surface of Mars. This type of optical illusion was demonstrated in a series of experiments [55], and subsequent observations showed that there were no canals on Mars.

This highlights another of the problems with the human eye as an astronomical detector. As well as being attached to an inappropriate processing system, the eye cannot store an image for examination by anybody other than the person who saw it. Early astronomers had to make sketches of their observations, and many of these would record the same perceptual illusions that were responsible for the canals of Mars. With only these sketches available, no matter how careful the observers might be, there was no objective recording of their observations that could be examined by others to check the results.

The human eye is also not very sensitive. One measure of the efficiency of a light detector is the fraction of photons received that are actually detected. This is known as detector quantum efficiency, or DQE. The greater this fraction the more photons we see, and the greater the sensitivity. An ideal detector would detect 100% of the photons falling onto it. Unfortunately the human eye does rather poorly, detecting only around one percent of the photons that it receives [167]. Worse than that, the 'exposure time', during which the human eye collects light for detection, is rather small, roughly $1/15^{th}$ of a second. If we were able to hold our eyes fixed, and collect photons for just a whole second, we would be able to see things 15 times fainter than we would otherwise be able to, but there is no way to change the exposure time on our eyes. In contrast, artificial detectors, such as photographic film or silicon detectors, can have whatever exposure time we want. This is critically important when it comes to collecting photons from the many astronomical objects that are too faint for the human eye to see.

2.3 ASTROPHOTOGRAPHY

The first great detector revolution in astronomy was the invention of photography. While still operating only in the optical, photography allowed long exposure times, the storage of images for later examination, and solved many other problems associated with the Mark 1 Eyeball. There is a huge literature describing astrophotography, with plentiful books and scientific reviews [41] covering the subject in detail. Photography was the main method of detection in astronomy for more than a century, and there are still large libraries of photographic plates stored at many of the world's observatories.

The process of light detection in a photographic plate is somewhat similar to that in the human eye - a photon triggers a chemical reaction. In the case of a photographic plate, though, the reaction involves silver halide, not proteins. When a photon of an appropriate wavelength interacts with a silver halide molecule, an atom of silver is produced. If further photons arrive at the same position, these silver atoms will accumulate into a stable speck of silver, producing what is known as a 'latent image'.

In astrophotography the story does not end at this point. After being exposed for a suitable length of time, the photographic plate is removed from the telescope and taken to a darkroom where it is developed. This is a chemical process whereby the silver halide on the photographic plate is turned into silver. However, where specks of silver are already present in the latent images, the change from silver halide to silver in the presence of the developer is much faster. The number of silver atoms in the latent images can increase by a factor of a billion or more, so the few atoms of silver in the latent image grow to become a readily discernible image on the photographic plate.

The efficiency of photographic plates is somewhat better than the human eye, reaching DQEs as high as 3% over long exposure times [95]. To reach this kind of sensitivity a series of operations known as hypersensitizing are required. This can involve a pre-sensitising flash of light, and/or exposing the photographic plate to vacuum, heat, or gaseous hydrogen and nitrogen [151]. From the point of view of an astronomer who has used electronic detectors for my whole career, many of these processes sound like black magic, but they were the mainstay of most of astronomy for a considerable period.

Photographic plates are inherently most sensitive to blue light, at wavelengths shorter than about 500 nm (0.5 microns). They can be made to work at longer wavelengths, out to about 800 nm, using dyes that fluoresce in the blue when exposed to longer wavelength light, but astrophotography cannot push much beyond the optical part of the spectrum. For sensitivity at longer wavelengths, astronomers had to await the development of detectors based on electronic devices rather than the wet chemistry of the eye or photographic plates.

2.4 MAKING ELECTRONS FROM PHOTONS

So far we have discussed the detection of photons in fairly complex chemical systems, be they the proteins in the eye, or the chain of chemical reactions in the photographic process. The way photons are detected in the semiconductor detectors used for most optical and infrared astronomy today is somewhat simpler, at least in concept. The technique goes back to the first experiments that demonstrated that light was made up of the discrete particles that we now call photons, and involves something called the photoelectric effect, and a device called a gold leaf electroscope.

An electroscope consists of a metal plate attached to a metal rod that leads inside a sealed chamber holding a vacuum. Inside the chamber a piece of gold leaf is attached to the metal rod at one point, allowing it to dangle vertically. A window in the chamber allows you to see what is going on inside.

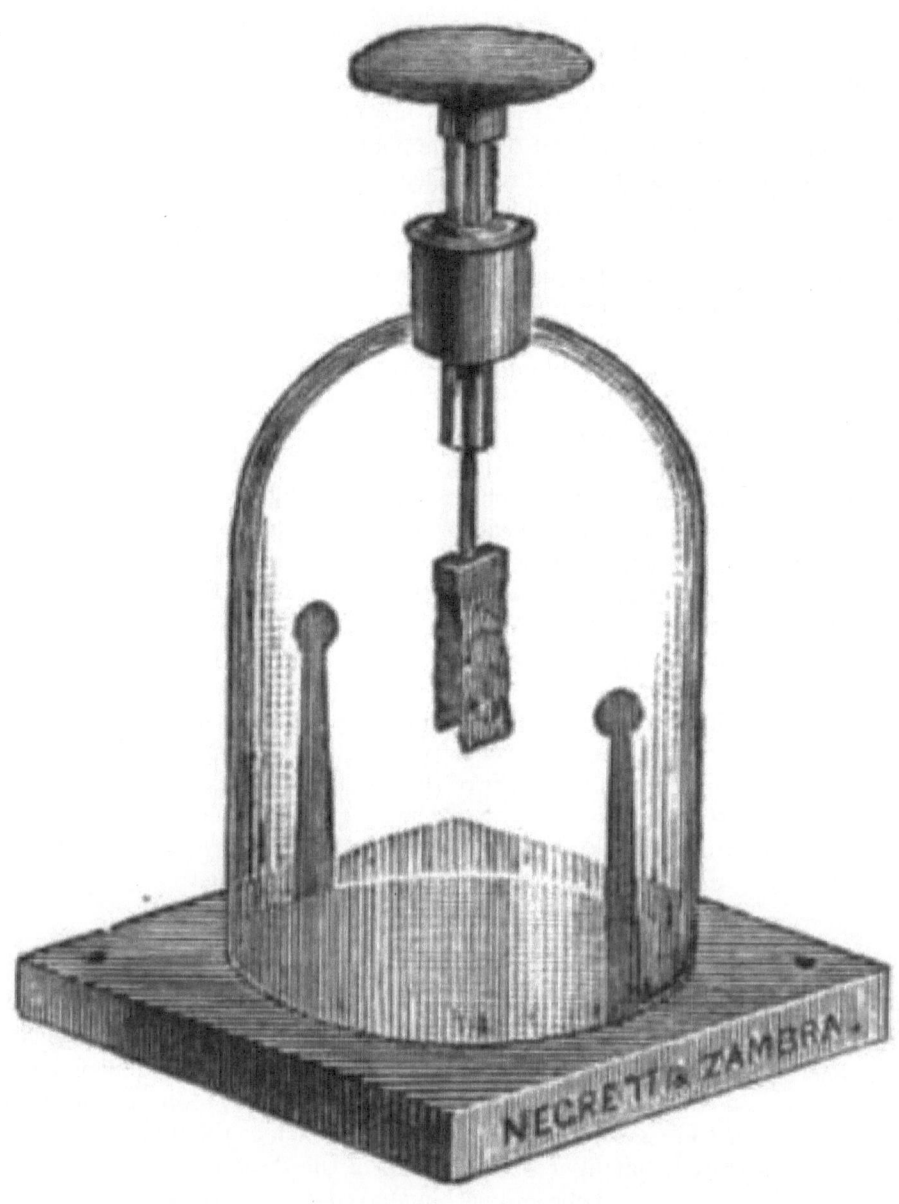

Figure 2.2 A 1910 diagram of a gold leaf electroscope [113].

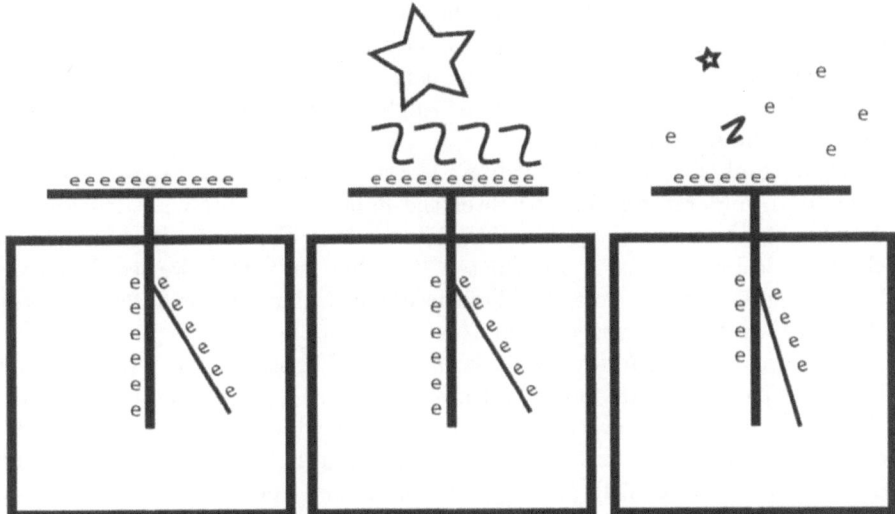

Figure 2.3 The operation of an electroscope.
Left: A charged-up electroscope, with electrons holding the gold leaf away from the central rod. Centre: A bright light shines on the top plate, but at too low a frequency to eject electrons. Right: A faint light, operating at a high enough frequency, will eject electrons and discharge the electroscope, no matter how faint the light might be.

A drawing of a 1910 electroscope is shown in Figure 2.2. If the device isn't otherwise interfered with, the gold leaf will simply lie alongside the vertical rod inside the vacuum chamber. Things change if the metal plate at the top of the electroscope is charged up with a source of static electricity, such as a plastic rod rubbed with wool. Static electricity is produced when electrons are knocked off atoms in the wool and collected on the plastic rod. When these electrons are discharged, you get the familiar cracking noise, flash of light and, if you're unlucky, a brief electric shock. When the source of static electricity is connected to the electroscope, electrons flow onto the metal, and cover everything electrically connected to it. This includes both the rod and the gold leaf. Since similar electrical charges repel each other, there is a small force between the metal rod and the gold leaf inside the electroscope, and the gold leaf no longer dangles vertically. Instead, it angles away from the metal rod, held there by the mutual repulsion between the electrons on the metal rod and those on the gold leaf itself (see Figure 2.3).

So what does this have to do with the detection of light, and the demonstration that light is made up of photons? The key thing is what happens when you shine light onto the metal plate at the top of the electroscope. Depending on the metal the plate is made from, and the frequency of the light that is shined on it, one of two things happens: Either the electrons are knocked off the plate by the light, discharging the electroscope so that the gold leaf can settle downwards; or nothing happens at all, no matter how bright the light might be. Bluer, higher frequency, light is able to knock electrons off a wider range of metals. The process by which light knocks electrons off the metal is known as the photoelectric effect.

If we think that light is simply an electromagnetic wave, these results are difficult to understand. The brighter the light, no matter what frequency it might be, the more energy is being delivered, so the easier it should be for electrons to be removed from a metal. But that isn't what we see. In fact, as long as the light is above a threshold frequency, different for different metals, the plate will be discharged no matter how faint the light is.

This was a problem for the classical view of light as a wave phenomenon, but is simple to understand if we think of light as being made up of separate particles, which we now call photons, each of which has an energy that is directly related to its frequency. The idea that light has both particle-like properties - that it is divided up into photons - and wave-like properties - each of these particles can be thought of having a wavelength - was first suggested by Einstein in 1905, and led to the award of the 1921 Nobel Prize for physics[1].

The direct use of the photoelectric effect to detect photons in astronomy started with photomultipliers in the 1970s. These devices start off with a photocathode, something very similar to the plate in an electroscope from which electrons escape when hit with light above a certain frequency. In the photomultiplier, though, these electrons are not simply allowed to escape. instead, they are accelerated by an electric field, and then allowed to hit a second surface. Since the electron has been accelerated, this second collision liberates many more electrons, and the process can be repeated through a series of accelerations and collisions, until finally these electrons are collected, and the initial photon that fell on the photocathode can be counted.

The total number of electrons detected in such a device can be as high as a million, and the detector quantum efficiency can be as high as 25%. This compares very well to the few % that the Mark 1 Eyeball and photographic plates can achieve, but it is still a long way from 100%. Also, photomultipliers are more sensitive to bluer light because of the energy needed to kick photons out of the photocathode. They thus are not very useful for infrared astronomy, but this kind of detector is still used for ultraviolet astronomy on board satellites such as GALEX [105].

2.5 THE DIGITAL AGE

The detectors used at most optical observatories today, and in the camera of your mobile phone, still rely on the photoelectric effect, but are not based

---

[1]Though he didn't actually receive this prize until 1922.

on photomultipliers. Instead of photocathodes, they use semiconductor light detectors known as charge coupled devices - CCDs.

A pure semiconductor such as silicon is a poor conductor of electricity since all the electrons in the material are closely tied to silicon atoms in what is called the valence band. Silicon has four electrons in its valence band that are responsible for binding each silicon atom together with its neighbours in a crystalline structure. If a small amount of an element which has five electrons in its valence band, such as arsenic, is mixed into the silicon crystal in a process called doping, then there are spare electrons available which can carry current. This is known as n-type doping since negative charge carriers - electrons - are being added to the semiconductor. Similarly, if the silicon is doped with an element, such as boron, that has three electrons rather than four in its valence band, you produce positively charged holes in the crystal. These holes can move around the semiconductor, producing what are effectively positively charged carriers of charge, so this process is called p-type doping.

Careful preparation of a semiconductor device with different layers of p- and n-type doping, and the application of a voltage across the device, can produce inside the silicon something that is very much like an electroscope - places where electrons can be trapped called capacitors. Incoming photons of an appropriate energy can knock electrons out of these capacitors, reducing the electric charge stored in them. At some later point, once the exposure is completed, the electric charge remaining in the capacitors can be measured. The fewer electrons there are, the more photons have been received by the detector. CCDs also have the advantage that the contents of the capacitors can be easily moved around. The device essentially acts as a 'bucket brigade', where the contents of one capacitor can be moved to any of its neighbours by a simple operation. If you combine this with the capability of producing large numbers of these detectors in a two-dimensional array, you can have a detector sensitive over a large area, which can also be read out very simply.

There are a number of other advantages to this kind of detector over photomultpliers. Firstly, silicon-based semiconductor devices allow a wider range of wavelengths to be detected, covering the whole of what astronomers refer to as the optical band, from about 360 nm (0.36 microns) to about 1000 nm (1 micron) in wavelength. Secondly, and more importantly, the quantum efficiency of CCDs can be very high. Some modern devices can have DQE values of 90% or higher. Finally, the fact that these are silicon devices means that all the technologies developed by the silicon chip industry can be brought to bear to make high-quality detectors at reasonable prices. Today, even amateur astronomers can have better CCD cameras on their telescopes at home than were being used by professionals at large observatories when I started my PhD in 1986. At the same time professional observatories now operate with CCD cameras that have many hundreds of millions of pixels. One of the largest such detectors, the DECam instrument for the Dark Energy Survey [144], is a 570 million pixel camera attached to a 4 m optical telescope. The camera itself, which can be seen in Figure 2.4, is made up of 62 separate CCD detectors.

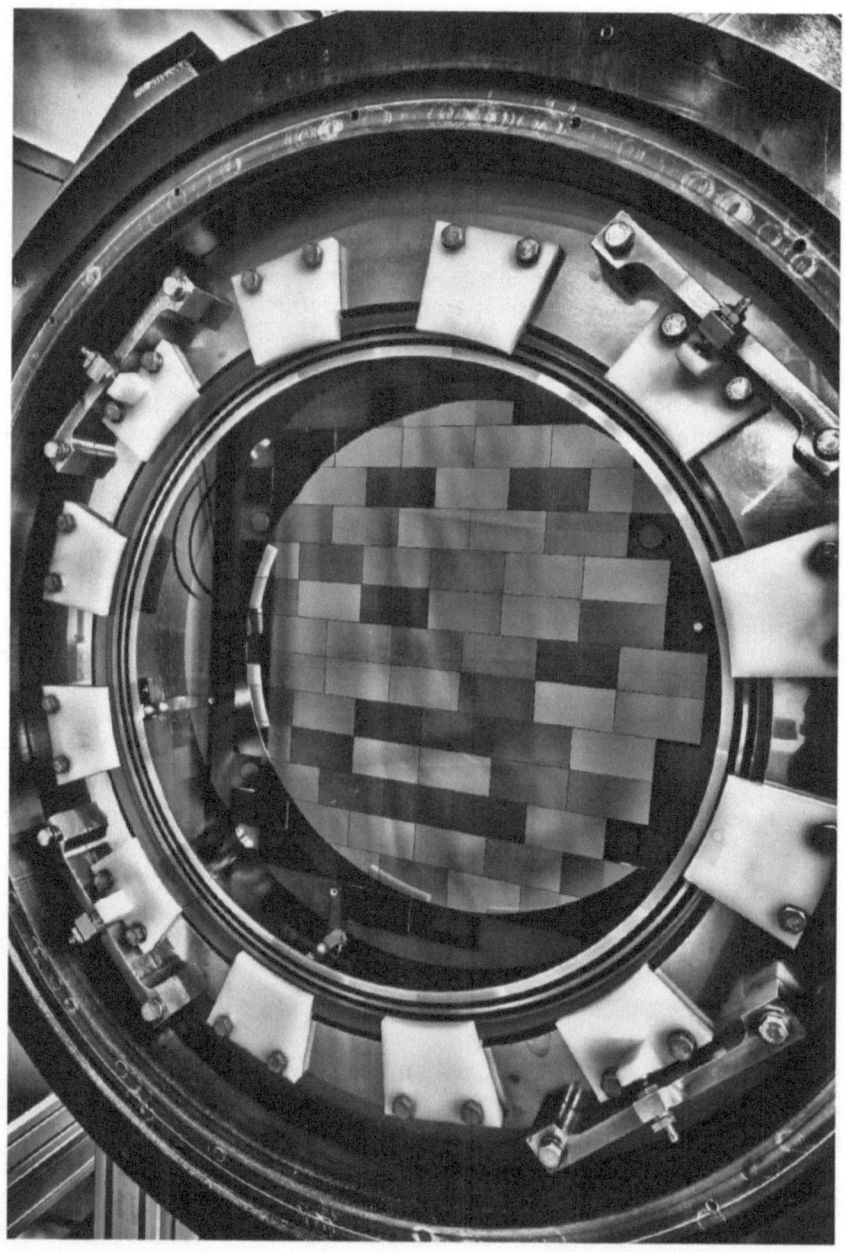

Figure 2.4 The DECam camera, to be used for the Dark Energy Survey at the 4 m Blanco Telescope in Chile.
DECam is a 570 million pixel camera made up of 62 individual CCD detectors, some of which can be seen in this picture. (Figure from http://www.darkenergysurvey.org/. Courtesy of Fermilab/DES.)

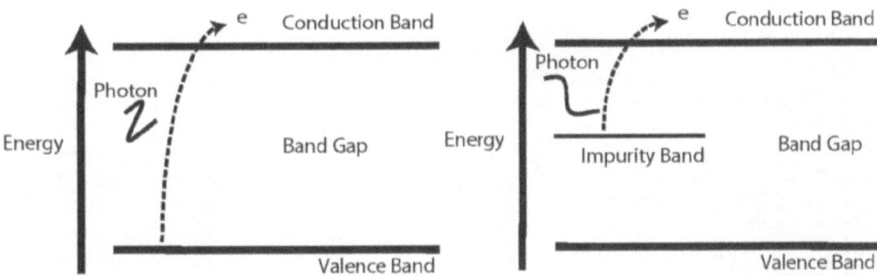

Figure 2.5 Impurities in a semiconductor.
Energy level diagrams for an intrinsic photoconductor (left) and for an extrinsic photoconductor where an impurity allows lower energy photons, and thus longer wavelength light, to be detected.

2.6 INTO THE INFRARED

The properties of silicon as a semiconductor mean that it is insensitive to radiation at wavelengths longer than about 1000 nm (1 micron). This means it cannot be used as a detector in the infrared, and many of the conveniences of CCD detectors are lost. Instead, for most wavelengths of infrared light, we can use devices called photoconductors based on more exotic semiconductors such as indium antimonide (InSb), mercury-cadmium-telluride (HgCdTe), silicon arsenide, Si:As (silicon doped with arsenic) or, at the longest wavelengths, germanium:gallium.

The idea behind these devices is fairly simple. In these semiconductors, as in silicon itself, electrons are either bound closely to atoms and are responsible for holding the material together in the valence band, or they can float freely through the material in the conduction band. The normal state for electrons in these materials is to sit in the valence band, but an incoming photon of an appropriate energy can be absorbed by a valence band electron, exciting it into the conduction band, and allowing electrical conductivity. See Figure 2.5.

The more photons that arrive, the more excited electrons end up in the conduction band, and the greater the electrical conductivity. Since there is an energy gap between the valence and conduction bands, only photons with an energy greater than this energy gap, and thus a frequency higher than the corresponding value, can be detected by the photoconductor. This is directly equivalent to the energy limit for photons knocking electrons off electroscope plates in the photoelectric effect but, since it takes place within a material, the energies involved are generally much smaller. This means we can use photoconductors to detect lower energy, infrared light.

Further tricks can be played to extend the range of wavelengths over which photoconductors can work. The standard way of doing this is to introduce impurities into the material which can, in effect, provide an intermediate,

Figure 2.6 Stressed infrared detectors.
The PACS detector subassemblies. The large curved piece of metal visible at the top of the assembly is the leaf spring responsible for stressing the detectors, extending their sensitivity to the long wavelengths at which this detector has to operate. (Figure used with permission from MPE and the PACS Consortium.)

stepping stone, energy level between the valence band of the raw semiconductor and the conduction band. Gallium doped germanium has been used in this way to make photoconductors sensitive out to about 100 microns in wavelength, right in the middle of the far-infrared.

To get to sensitivity at still longer wavelengths in photoconductors requires extreme measures. The PACS instrument on the Herschel Space Observatory has Ge:Ga photoconductors sensitive out to about 200 microns. They can reach wavelengths as long as this only because the detectors are stressed (see Figure 2.6), effectively pushing all the energy levels closer together. A large leaf spring is used to do this for the PACS detectors, which is not something you would expect to have in a high precision astronomical detector. Needless to say, such extreme detector engineering is very delicate.

Astronomy would be very slow if it were done one single photoconductor detector at a time. If we want to produce images of the sky, or measure the spectra of specific objects, in a reasonable amount of time, we need large arrays of detectors whose results can be read out rapidly. There is no 'bucket chain' way of moving electrons around on a photoconductor the way you can with a CCD, so, for many astronomical detectors, this readout task is subcontracted to a CCD. The 'buckets', or capacitors, that hold electrons in a CCD don't care where the electrons come from. They could come from another device, such as a photoconductor, that is electrically bonded to the CCD pixel. This is exactly what is done, for example, in the detector arrays of the IRAC infrared camera on the Spitzer Space Telescope. The InSb and Si:As photoconductors in the camera detect the infrared photons, then pass the electrons they produce onto a CCD pixel that is 'bump bonded' to the photoconductors by a small dot of indium metal.

A lot of the basic technology development for both CCDs and infrared detectors was done by astronomers and engineers working for observatories, but not all of it. The availability of sensitive, high-speed, mid-IR detectors is also of great interest to the military, who need to see things that are warm, see through obscuring smoke and see things at night. That is why some of the detector arrays of the earliest infrared cameras at astronomical observatories have the words 'tank buster' etched onto their packaging - they originally came from anti-tank missiles. This link with the military continues to the present day. The exquisitely sensitive mid-infrared detectors in the MIRI instrument (Mid InfraRed Instrument) for the James Webb Space Telescope were developed using top secret military techniques, and are thus deemed to be 'strategic materials' requiring an interesting and complex set of security clearances for their export and use since MIRI is a European Instrument, but the detectors come from the US.

2.7 LONGER WAVELENGTHS

Pushing photoconductors to work at ever longer wavelengths, through precision doping with impurities and detector stressing, allows us to detect photons out to wavelengths of ~ 200 microns, but such extreme measures cannot be pushed to much longer wavelengths. Even at 200 microns some of the side effects of these methods can be quite inconvenient. These detectors can have memory effects, delayed responses, non-linear reactions and other complicating factors that make them difficult to use. So how can we go about detecting light at even longer wavelengths, into the submillimetre part of the electromagnetic spectrum?

For this we need to use a completely different detector technology: *bolometers*.

In concept, a bolometer detector is very simple, and not that different to the differential thermometry that William Herschel used to first detect infrared light. In a bolometer detector, light is absorbed by some absorber, for instance sapphire or amorphous silicon. The energy of the absorbed light causes the absorber to heat up a little, and a thermistor measures this temperature rise. A diagram of how a bolometer functions can be found in Figure 2.7. However, the

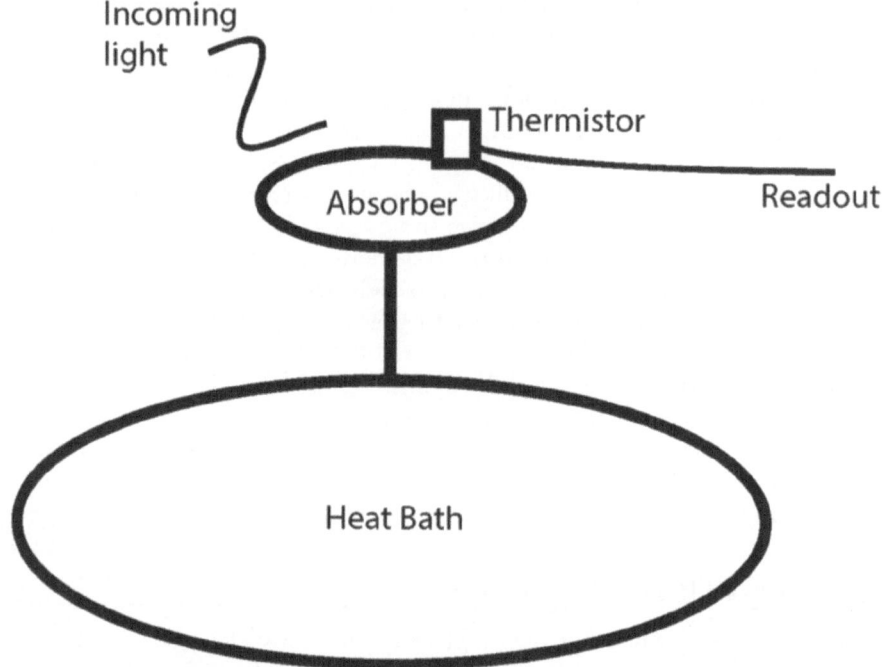

Figure 2.7 How a bolometer functions.
Incoming radiation is absorbed by the absorber. The energy from this radiation heats the absorber up, and this temperature change is detected by the thermistor. The temperature of the bolometer is kept stable in the long term through connection to a cryogenic heat bath, typically only just above absolute zero.

power received during the detection of far-infrared and submillimetre radiation from astronomical sources is very small. For this to produce a detectable rise in temperature in a bolometer you need an absorber with a very low heat capacity, to maximise the temperature rise from a small amount of received power, and a thermistor that is very sensitive. Both of these requirements can only be reached by operating at very low temperatures, just a few tenths of a degree above absolute zero, -273 C.

At these very low temperatures, most materials have a very high temperature coefficient of electrical resistivity. This means that their electrical resistance changes a lot even for a very small change in temperature. The thermistors on bolometer instruments such as the SPIRE instrument on the Herschel Space Observatory [70] rely on this simple principle, and are made out of the metal germanium. They are sufficiently sensitive that just a hundred far-infrared photons are enough to produce a detectable change in temperature.

More modern bolometer systems, such as the SCUBA2 instrument [78] currently working on the James Clerk Maxwell Telescope (JCMT), take advantage of superconductivity to boost the sensitivity still further. Superconductivity is the property of some materials at very low temperatures that allows them to conduct electricity with no electrical resistance whatsoever. Superconductivity kicks in at a specific temperature for any given material. The detectors in SCUBA2 work by keeping their thermistors just below the temperature at which they become superconducting. Any small increase in temperature due to incident radiation will push them over the superconductor-non-superconductor transition, and their electrical resistivity will rise massively. These Transition Edge Superconducting (TES) bolometers can also be manufactured in large arrays more easily than the, essentially hand-made, individual devices that are used in SPIRE, allowing, for the first time, the kind of large area detectors in the submillimetre part of the spectrum that astronomers have got used to in the optical and near-infrared.

In principle, since all a bolometer is detecting is a rise in temperature, they can be used to detect anything that might deposit energy in them. This is both an advantage and a disadvantage. It is an advantage because they are very flexible detectors. As well as long wavelength submm radiation, they have also been used to detect X-rays, and to measure the energy of each arriving X-ray photon with far greater precision than other detector technologies. Bolometers can also detect gamma rays, and measure their energy with precision, opening up the possibility of using bolometers to detect any attempts to smuggle plutonium, a key constituent of nuclear weapons, through ports or airports. The capability of the SCUBA2 TES bolometers as plutonium detectors was tested while they were being assembled. Unfortunately something went a little wrong, and the lab where these tests were being made got contaminated with a small amount of plutonium, necessitating a lengthy decontamination process. The SCUBA2 detectors were sealed inside this lab until it was decontaminated, delaying their delivery to the instrument builders, and the completion of the project, by several months.

The disadvantage of bolometers is exactly the same thing that makes them so flexible - their ability to detect anything depositing energy in the absorber. This is another reason why they have to be kept very cold, with all the complications that implies. You also have to make sure they are shielded from any other source of radiation that might get into the instrument. This problem applies at some level to all the detector technologies we've looked at. A long-standing, but unavoidable annoyance, of CCD images is that they will always contain a number of bright pixels produced by cosmic rays that have plunged through the Earth's atmosphere and have blundered into your astronomical detector. The effects of cosmic rays can be removed with various post processing techniques and observing strategies, but they are our first example of the practical obstacles that get in the way of infrared astronomy.

2.8   INTO THE RADIO

At wavelengths longer than ∼1mm, the techniques of radio astronomy become more and more applicable. These are not the subject of this book. Unlike the approaches outlined above, they are far less concerned with the quantum nature of light - radio astronomers don't say too much about photons. Instead, the techniques used at wavelengths longer than about 1mm are based on the wave nature of light, directly measuring the rise and fall of the electric field associated with electromagnetic radiation, and manipulating these waves.

This kind of approach, termed coherent detection, is in contrast to the incoherent detection approaches discussed above. There are some advantages to this. You can build large arrays of radio telescopes spread over large distances, record the variation of the electromagnetic waves being received, and then combine them at a later date. This process, known as interferometry, can produce very high resolution images by effectively having the resolving power of a telescope as big as the longest separation between the individual radio dishes. That distance can be as large as the diameter of the Earth, or, if radio telescopes are launched into space, even larger.

These coherent detection techniques are gradually being expanded to higher and higher frequencies, allowing them to start being used in the submm and far-infrared. Projects, such as the Submillimetre Array (SMA) and Atacama Large Millimetre Array (ALMA), are starting to bring these kinds of capabilities to the longest wavelengths we'll be examining in this book.

2.9   THE ENEMY: WHAT GETS IN THE WAY

While we now have very capable detectors working all the way from the near-infrared, at wavelengths of just a few microns, to the far-infrared with wavelengths of a few hundred microns and beyond, everything is not plain sailing. Being able to detect radiation at these wavelengths is only half the job, be-

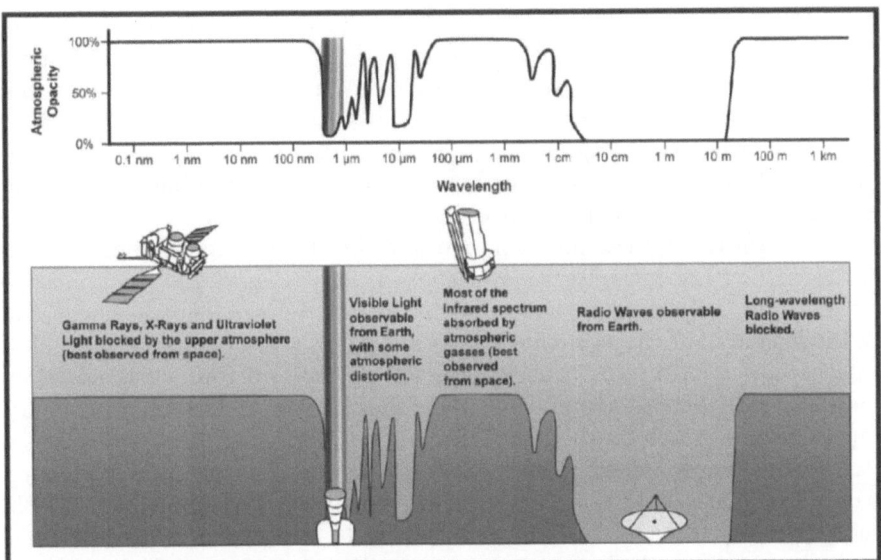

Figure 2.8 Atmospheric opacity.
The opacity of the Earth's atmosphere at different wavelengths, from gamma
rays to radio. 100% opacity means that nothing that hits the top of the atmosphere can reach the surface of the Earth, so ground based astronomy at,
for example, ultraviolet wavelengths, is impossible. (Courtesy of NASA.)

cause there are many things that get in the way of our observations. Escaping
these obstacles means, in some cases, that we must take extreme measures,
such as sending our observatories into space, but even ground-based infrared
telescopes have a harder time than their optical counterparts.

One of the key difficulties facing infrared astronomers is the atmosphere,
since atmospheric transparency varies considerably from one wavelength to
another. This is shown in Figure 2.8.

The two most transparent parts of the atmosphere are at radio wavelengths
and in the optical. The latter is rather useful since it means that we can use
our eyes to see things far away, and, in the case of astronomy, to see right
through the atmosphere and beyond, to distant stars and galaxies. If you
combine the transparency of the atmosphere at optical wavelengths with the
fact that the thermal black body spectrum of emission from the Sun peaks at
these wavelengths as well, then it's clear why our eyes evolved to work in this
part of the electromagnetic spectrum.

Looking beyond the optical to the shorter wavelengths, and higher frequencies, of the ultraviolet, X-rays and gamma rays, you'll see that the atmosphere
is fully opaque at these wavelengths. This is a good thing, since the high energy photons at these wavelengths would be very damaging if we were exposed

to them all day every day. However, this does mean that anybody interested in astronomy at these wavelengths cannot operate from the ground. Instead they have to launch their telescopes into space. At the very longest wavelengths, beyond normal radio frequencies, the atmosphere is also opaque thanks to effects in the Earth's ionosphere. Not much astronomy has been done at frequencies below 30 MHz, but there are plans to follow the high energy X-ray and gamma ray astronomers into space.

Between the optical and radio parts of the electromagnetic spectrum lies, as we know, the infrared. As you can see from Figure 2.8 the atmosphere ranges from reasonable transparency in the near-infrared, at wavelengths just a bit longer than those in the optical, to completely opaque by the time we reach the far-infrared at wavelengths around 100 microns. The situation is actually rather more complicated than that. If you look at the atmospheric opacity more closely in the near-infrared, you'll see that there are spikes and troughs of transparency. This means that at some wavelengths you can see through the atmosphere almost as well as you can in the optical, but at other, not that very different, wavelengths, trying to look through the atmosphere is like trying to look through a brick wall. Because of this effect, we have to start talking about atmospheric 'windows'.

Figure 2.9 shows the atmospheric transparency of part of the near-infrared region, covering wavelengths from 1 to 2.6 microns. There is a lot of structure to the atmospheric transparency over these wavelengths. This is because most of the absorption at these wavelengths comes from water molecules, which have a large number of transitions in this region. What we're effectively seeing in Figure 2.9 is the result of a lot of individual absorption lines all blending together. The end result is that there are three parts of this region of the near-infrared where there is reasonably good transmission, centred at about 1.2, 1.6 and 2.2 microns, and the atmosphere is largely opaque between them. This gives us the concept of an atmospheric window - a contiguous range of wavelengths where the atmosphere is reasonably transparent. The three windows present in the range of wavelengths shown here are named J, H and K[2].

There are further windows of ever-decreasing quality up to a wavelength of 20 microns. At this point we are into the mid-infrared. Beyond 20 microns, absorption from atmospheric water vapour, and other species such as carbon dioxide and oxygen, gets too much, and the atmosphere becomes completely opaque. The next windows start to open at wavelengths of a few hundred microns, though some of them are still so poor that they might be better

---

[2]The names given to these bands is a little eccentric. Originally it was meant to be a simple alphabetic extension from the longest wavelength optical filter defined at the time that near-infrared astronomy was getting started. That filter was called I, so the next bands to be named, in the infrared, were J and K. However, once astronomers started going up mountains, they realised that there was an additional window, centred on 1.6 microns, which was viable for astronomical use. This was named H, so the infrared astronomer's alphabet goes I, J, H, K. Things are now rather more complicated, with two extra filters slipped in between I and J called Z and Y (in order of increasing wavelength).

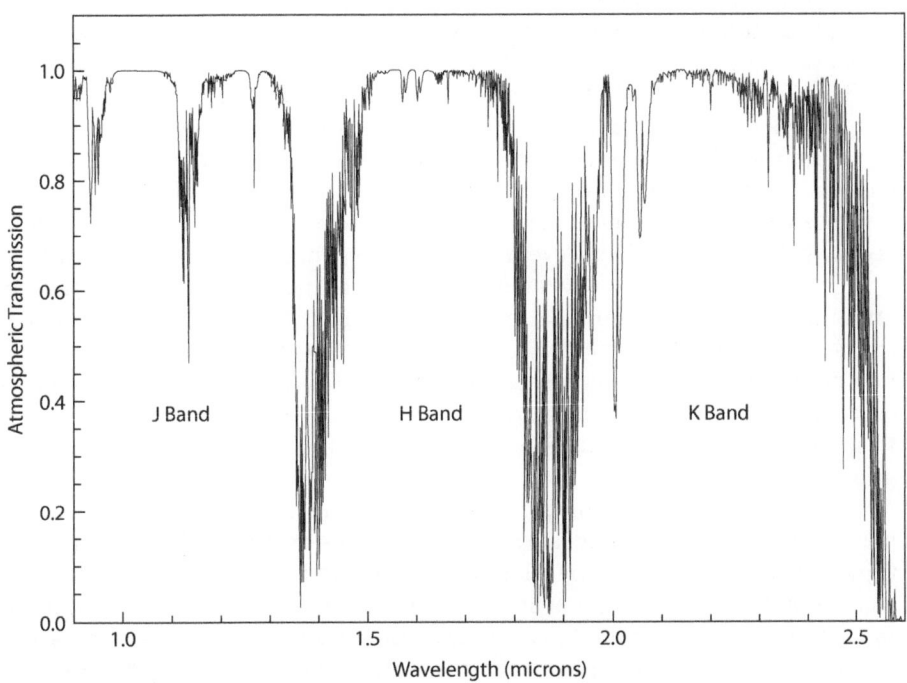

Figure 2.9 Near-infrared windows.
Atmospheric transmission in the near-infrared at the Mauna Kea observatory, showing the J, H and K bands and their respective atmospheric windows. (Figure created from data provided by Tom Geballe and the Gemini Observatory.)

described as doors than windows. These get increasingly more transparent (see Figure 2.8) until we reach the radio.

2.10 DEFEATING THE ENEMY: CLIMB A MOUNTAIN

Atmospheric absorption is difficult and expensive to avoid completely. For some wavelengths, where the atmosphere is opaque - the far-infrared, where Herschel operates, as well as for gamma-rays, X-rays and UV - there is no real alternative to operating in space. But there are some half-way houses that have allowed infrared astronomy to develop without the help of space vehicles.

The first of these methods is to move to a high and dry location, such as the tops of mountains, or high altitude plateaus in mountain ranges. The higher you go, the thinner the air gets and, as a result, the amount of absorbing material between you and the stars is reduced. At the Mauna Kea observatory in Hawaii, for example (see Figure 2.10), the telescopes are at an altitude of about 4200 m. This places them, roughly speaking, above about a third of the Earth's atmosphere, and allows some of the poorer windows in the far-infrared and submillimetre parts of the spectrum to become usable. Going to these locations also helps with more general weather issues. You can still lose entire nights, or whole observing runs, to thick clouds or snow, but the chances of this happening are greatly reduced since the worst weather tends to go around this mountain in the middle of the Pacific ocean rather than over it.

The need to select high, dry and, ideally, cold sites limits the number of places where you can build an astronomical observatory. The best sites are typically on isolated volcanic islands, like the Big Island of Hawaii in the Pacific, or on the Canary Islands in the Atlantic. Alternatively, sites in mountain ranges near to coasts, whose weather is dominated by cold ocean currents, also provide good conditions. This is why many observatories are located on a variety of peaks in the Chilean Andes or in the California mountains. The cold ocean currents, the Humboldt Current in the case of Chile, reduce the amount of rainfall inland. This leads, for example, to the extremely arid nature of the Atacama desert in Chile, which is home to the European Southern Observatory's (ESO) Very Large Telescope (VLT), Atacama Large Millimetre Array (ALMA) and other associated telescopes. The Atacama is also the site of the planned European Extremely Large Telescope (E-ELT).

The high altitudes and extremely dry conditions at these observatories are great for infrared astronomy, but they do cause problems for the observers. The altitude can lead to altitude sickness and potentially life-threatening conditions such as pulmonary or cerebral oedema, while the dryness can lead to nosebleeds and other problems. In some cases, telescope control rooms have been equipped to provide an oxygen-enriched atmosphere, while in others those working at the telescopes are given personal oxygen supplies.

An interesting side effect of the reduced oxygen at these high altitude

Figure 2.10 Three of the observatories on the summit of Mauna Kea. From left to right they are: Subaru, the Japanese 8 m telescope, the two 10 m telescopes of the Keck Observatory, and the 3 m telescope of the NASA Infrared Telescope Facility (IRTF). All of these telescopes can observe both the near- and mid-infrared.

observatories is that the eye is less sensitive. What you can see by eye at these supposedly world-class observatory sites is often less impressive than you might expect. However, if you are lucky enough to have your own oxygen supply, then you can correct this situation. I'm told that the sky above Mauna Kea is as impressive as you would expect for a world class observatory once you can see it properly. Sadly the observatories I have used on Mauna Kea reserve their oxygen supplies for emergencies, so this is something I have yet to see for myself.

2.11   FLY A PLANE, FLOAT A BALLOON, BUILD A ROCKET

There is quite a gap between the highest mountains and Earth orbit, from altitudes of a few thousand metres to orbits of a few hundred kilometres. It is possible, though quite difficult, to deploy telescopes in this gap. There are two ways to achieve this. Altitudes up to 14000 m can be reached by aircraft equipped with telescopes such as the Kuiper Airborne Observatory (KAO) or, more recently, the Stratospheric Observatory for Infrared Astronomy (SOFIA). The latter is essentially a 747 Jumbo Jet with a hole cut into it and a 2.7 m diameter telescope pointing out - see Figure 2.11.

Still higher altitudes, and clearer skies, can be reached by helium balloons. These are not the small party balloons that you might be used to, but much larger and stronger. The balloon that flew BLAST (the Balloon-borne Large Aperture Submillimetre Telescope), a test bed for the SPIRE instrument aboard Herschel, was about 200 m in diameter, and contained about two million cubic metres of helium. Astronomical balloons typically fly at an altitude of about 40 km.

While planes and especially balloons can fly above a lot of the absorption in the atmosphere, their flights are by their very nature short in duration. A typical observing flight on SOFIA lasts about 16 hours, while the longest duration balloon flights are about two weeks. While SOFIA can, in principle, fly many flights each year, balloon payloads are notorious for requiring reconstruction after descending back to the ground. The first long duration flight of BLAST, for example, ended up with the payload, including the instrument and telescope, breaking up and being scattered along a 200 km debris trail when the descent parachutes failed to detach after landing. The observational data, though, was recovered intact and the flight was a great success.

The best way of avoiding all the problems associated with the Earth's atmosphere is, of course, to leave it far behind, which is why we have space-based astronomy in the infrared. Space observatories with some kind of infrared capability have been flying since the early 1980s, starting off with IRAS, the Infrared Astronomy Satellite, and continuing with ISO, the Infrared Space Observatory, in the 1990s, Spitzer and Akari in the 2000s, and continuing with Herschel and, to some extent, Planck, which were launched together in 2009. The Hubble Space Telescope (HST) also has some capabilities in the near-infrared, while its planned successor, the James Webb Space Telescope

Figure 2.11 The SOFIA aircraft.
SOFIA, with its 2.7 m telescope visible at the rear of the plane. (Courtesy of NASA/ SOFIA.)

(JWST) will operate in the near- and mid-infrared. This, and other planned missions, will be discussed in Chapter 9, while details of the older observatories were covered in Chapter 1.

2.12   THE OTHER ENEMY: BACKGROUNDS

While the absorption of the atmosphere is the primary obstacle to infrared astronomy at many wavelengths, it isn't the only problem. One of the other difficulties comes because we live in a warm environment. This makes things nice for us but, as you will remember from our discussion of black body thermal radiation in Chapter 1, it means that everything around us gives off infrared light - infrared light is heat radiation, after all. This provides a local background signal which can swamp what you are looking for from an astronomical source.

Background isn't so much of a problem in the near-infrared, but by the time you reach the mid-infrared, at wavelengths around 10 microns, you have reached the point at which the black body radiation from something at room temperature will far outshine the brightness of any astronomical source other than the Sun. The effect is rather like trying to do optical astronomy during daytime[3]. In fact it's rather worse, since everything in the telescope environment - building, superstructure, the mirrors, the electronics, and the astronomers working on the telescope - will be glowing, and will, in principle, vary in the amount of infrared radiation they are emitting (astronomers, for example, who for these purposes act like large, bright, infrared floodlights, might walk around the building).

The way the effects of backgrounds are dealt with is to take very careful measurements of the background emission and subtract this signal from the astronomical signal you are looking for. The precision with which this needs to be done increases as you try to observe at longer and longer infrared wavelengths. The speed with which the background varies also increases at longer wavelengths. For ground-based submillimetre observations a technique called 'chopping-and-nodding' is used, where you switch your observations very quickly from your target source to a nearby piece of (hopefully) blank sky. The signals from these two positions are subtracted, hopefully leaving you with just the signal from your target.

The other way to escape the problems of local thermal backgrounds is, once again, to go into space, where large parts of your telescope and its surroundings can be cooled to very low temperatures. Many infrared observatories, including IRAS, ISO, Spitzer and Akari, have had their telescopes cooled to just a few degrees above absolute zero by liquid helium. This pretty much eliminates the entire thermal background. But there are problems with this approach. Firstly the liquid helium will eventually run out, limiting the lifespan

---

[3]Optical observing during the daytime isn't actually impossible. It is just, like observing in the mid-IR, very difficult. Unfortunately, the mid-infrared backgrounds don't conveniently go away at night.

of the satellite. Secondly, the large tanks of liquid helium needed for cooling that encircle the telescope limit its size. Spitzer, one of the largest telescopes with a cryogenically cooled primary, had a primary mirror just 85 cm across, something that would be considered distinctly undersized by ground-based professional astronomers.

The Herschel Space Observatory took a different approach, and instead of cooling its primary mirror with liquid helium, instead placed its mirror behind a large sunshade, and chose an orbit where this would keep the mirror perpetually shaded from the light of both the Sun and the Earth. The telescope then radiated its heat away into space until it eventually reached a temperature of about 90 degrees above absolute zero (about -183 C). At this temperature there was an equilibrium between the residual heat the mirror received, from the back side of the sunshade and waste heat from the spacecraft, and the heat it was radiating away into space. This allowed it to have a 3.5 m diameter primary mirror, much larger than any astronomical telescope ever put into orbit. You can see a picture of Herschel in Figure 7.10, and results of this mission will appear in much of the rest of the book.

2.13   CONCLUSIONS

In this chapter we have looked at how light and infrared radiation is detected from astronomical objects, and how the development of detector technology has allowed us to make ever better observations in the infrared. We have also seen the practical problems that the atmosphere and local backgrounds cause for these observations, and how this has led infrared astronomers to desolate mountaintops and into space.

The rest of this book is concerned not with detector technologies or observing techniques, but with what infrared astronomers found once we could make observations. In the next chapter we will start our survey of the infrared sky by looking at what infrared astronomy has told us about our nearest neighbours in astronomical terms - the planets, moons and other bodies that make up our Solar System.

CHAPTER 3

Local Heat

3.1 THE GREAT COLLISION OF 1994

Observing time on major telescopes is hard to get. You apply for time months, or years, in advance, and then your proposal gets judged against all the others that have come in. If you're lucky, you get allocated a few nights on some specific future dates on which to make your observations. You then refine your plans, book your plane tickets, and hope that bad weather or some other complicating factor doesn't mess everything up.

But the universe doesn't run to observatories' timetables, and sometimes something happens unexpectedly that is so unusual and important you just have to observe it no matter what other plans might be in place at the telescope. In the summer of 1994, I was meant to be making near-infrared observations of colliding galaxies when I was pre-empted by spectacular events in our own Solar System.

A little over a year earlier, an unusual comet was discovered and, as is the tradition with such things, was given the name Shoemaker-Levy 9 (SL9), as it was the ninth periodic comet discovered by the husband and wife team of Carolyn & Eugene Shoemaker, along with David Levy. The comet was unusual in a number of ways. Firstly, images showed that it was made up of multiple nuclei spread out over an elongated region. Secondly, it was unusually close to the planet Jupiter.

This wasn't a coincidence. It turned out that SL9 had passed close to Jupiter a few years before, so close that it was both captured into an orbit around the giant planet, and split apart by tidal forces during its closest approach. This interaction with Jupiter also doomed the comet to an eventual collision with the planet. And it was just my luck that the collision was due to happen during my observing run.

Its not often that you get a ringside seat for a cosmic cataclysm, so I was quite happy for the comet experts to take over the Calar Alto 3.5 m telescope I was meant to be using when Jupiter was available.

I arrived too late for the actual moment of collision, when the Calar Alto telescope was one of the first on the planet to see the impacts. Telephones rang as news agencies around the world sought a running commentary on the night's events. Several doom mongers had even predicted that a vast fireball would emerge from the impact, and spread across the Solar System to engulf the Earth and bring about the end of the world.

That did not, as you might guess, actually happen, but the series of impacts, as the multiple pieces of the shattered comet plunged one by one into the clouds of the gas giant's atmosphere, were certainly impressive. The observers saw bright fireballs in the near-infrared. The scars left by the impacts could be seen in Jupiter's atmosphere for months afterwards. My job at the telescope while this data was being taken was simply to go 'ooh' and 'ah' at the amazing images. It was quite a letdown to be allowed back to my own observations once Jupiter had set, and normal operations could get under way.

3.2 LOCAL GEOGRAPHY: A TOUR OF THE SOLAR SYSTEM

The Solar System is our local neighbourhood. It contains places we might physically visit with astronauts or robot proxies, and objects we can study in far greater detail than any that might be found elsewhere. And yet, since it is just one example Solar System compared to the tens or hundreds of billions of other solar systems in our own Galaxy, we have to be careful not to generalise too much from this single, special case.

The story of planetary astronomy since the second world war is one of transformation, of taking objects that had been, for the whole of human history until that point, only vaguely understood moving lights in the sky, and turning them into places, with their own geographies, place names and even well-recognised landmarks. Thanks to the robotic rovers now moving around on the surfaces of Mars and the Moon, you can now get landscape images from a different planet that are at least the equal of photographic landscapes of our own planet taken a hundred years ago. We even have images from the surface of the planet Venus and the moon Titan.

What, then, is our local neighbourhood like?

The centre of our Solar System is the Sun, our nearest star, and the source of the heat and light that keeps us alive. As we will see in the next chapter, the Sun is a normal, Main Sequence star, similar to many others in our Galaxy.

The first planet out from the Sun is Mercury. Its surface is baking hot when exposed to the Sun, up to 740K (about 470 degrees Celsius), but, thanks to the lack of an atmosphere, it is also freezing cold when out of the Sun, reaching temperatures as low as 80K. The heat of the Sun, combined with Mercury's low mass, led to any atmosphere it might have had as a young planet gradually leaking away into space. The lack of an atmosphere also means that its surface has no protection from asteroid impacts, so it is pockmarked by craters. Mercury is currently being studied in detail by the orbiting Messenger satellite, a NASA mission, which will be joined by ESA's Bepi/Colombo, due to be launched in 2016 and to arrive in 2024.

Next outwards from the Sun is Venus. In some ways Venus is a twin to the Earth, since they are roughly the same size and mass, but they have turned out very different. While the Earth has what we would think of as a pleasant and largely comfortable environment, the surface of Venus, at 735 K, is nearly as hot as that of Mercury, and the thick, heavy atmosphere produces surface pressures nearly a hundred times that of the Earth. The atmosphere is almost entirely made up of carbon dioxide, unlike the nitrogen and oxygen atmosphere of Earth, but there is sufficient sulphur dioxide to produce thick high level clouds that contain sulphuric acid droplets. These same clouds make it impossible for optical observations to see the surface of Venus. Heroic efforts over a number of years by the Soviet space programme eventually led to two spacecraft, Venera 9 and 10, successfully landing on the surface in 1975, transmitting the first-ever pictures from the surface of another planet. The landers didn't last long though, surviving just over an hour before succumbing to the hellish temperatures. Most of what we know about the surface of Venus comes from radar observations that can pierce the thick clouds, many of which were conducted from orbit by NASA's Magellan spacecraft.

Earth, our home, is the third planet out from the Sun. It is the only place in the Solar System to host life that we are aware of, and has a unique combination of liquid water on the surface and oxygen in its atmosphere. It is orbited by the Moon, making it exceptional among the other rocky, terrestrial planets. Mercury and Venus lack moons, while Mars, the fourth planet out from the Sun, has to make do with two tiny moons, Phobos and Deimos, which are likely just captured asteroids. The Earth's Moon is thought to have been produced in a cataclysmic collision during the early stages of the Solar System, between the young Earth and another young planetoid, possibly about the size of Mars.

The last rocky, terrestrial, planet in the Solar System is Mars, fourth planet out from the Sun. Like Venus, its atmosphere is dominated by carbon dioxide, but at a much lower pressure - less than a hundredth that of Earth. Its surface is cold, with temperatures ranging from a balmy 300K (roughly room temperature) at its hottest, to a bitter 130 K (-143 C) at the polar caps in winter. Mars has been subject to detailed exploration by a range of space probes that both orbited the planet and landed on its surface. These include the Mariner probes of the 1960s, the Viking landers of the 1970s through to the Spirit, Opportunity and Curiosity rovers currently operating there at the time of writing. Of all the planets in the Solar System, Mars is the best studied. The main reason for this is that Mars is one of the most likely places in the Solar System to have had, and maybe still have, liquid water flowing on its surface. We believe that liquid water is a requirement for the evolution of life, which makes Mars of great interest to us all, not just to astronomers. We will look at what infrared astronomy has told us about the possibilities of life

Figure 3.1 One of the first-ever landscape panoramas from the surface of Mars.
This image was captured by the Viking 1 lander, some of which can be seen in the foreground. The image was taken on 20th July 1976. (Courtesy of NASA.)

on Mars later in this chapter. Whole books have been, and are being, written on the results of our studies of Mars, so I will only be able to scratch the surface of this work here. The detailed studies of Mars over the last half century mean that we know its landscapes (see Figure 3.1 for example), geography, geology, places and weather rather well. In fact, thanks to the many orbiting satellites, it is likely that weather forecasts for Mars today are more reliable than weather forecasts for our own planet less than a century ago.

Jupiter is the next planet out from the Sun, but between Mars and Jupiter lies the asteroid belt. This is a collection of small rocky bodies, the largest of which, Ceres, has a diameter of 950 km and qualifies as a minor planet. While there are millions of asteroids, the total mass of material in the asteroid belt is actually very small, amounting to about 3×10^{21}kg, which is only 4% of the mass of the Moon. This material is spread over a huge volume, so the asteroid belt is actually very empty.

The asteroid belt owes its existence to the vast mass of the planet Jupiter, the influence of which, during the early stages of the Solar System, prevented the material between Mars and Jupiter from clumping together under gravity and forming a planet. Instead, most of the protoplanetary material lying in this region was nudged into unstable orbits by Jupiter, and either escaped the Solar System or fell inwards, towards a fiery death in collision with one of the terrestrial planets or with the Sun itself.

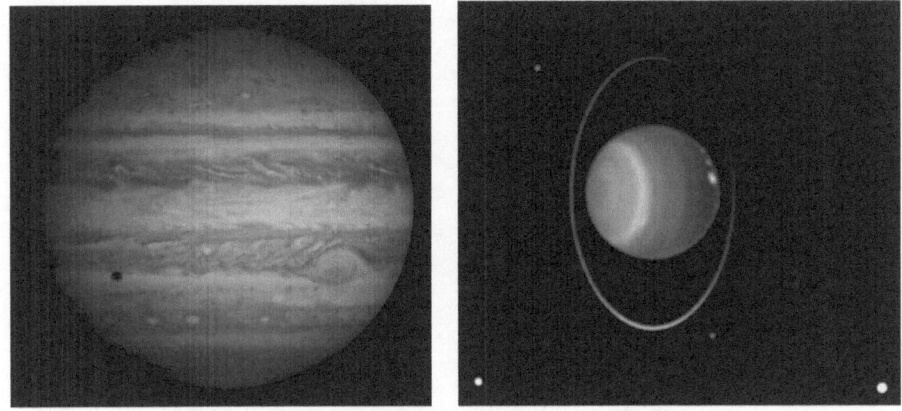

Figure 3.2 Jupiter and Uranus.
Left: This image was obtained by the Cassini spacecraft as it flew by the planet en route to Saturn in December 2000. The great red spot is visible towards the bottom right of the image. The black spot towards the bottom left is the shadow cast by Jupiter's moon Europa. Credit: NASA/JPL/University of Arizona. Right: Near-infrared image of Uranus obtained by the Hubble Space Telescope. This shows the banded structure of its cloud systems and some individual clouds, as well as the rings and some of Uranus' moons. (Left: Courtesy of NASA/JPL/University of Arizona. Right: Courtesy of NASA.)

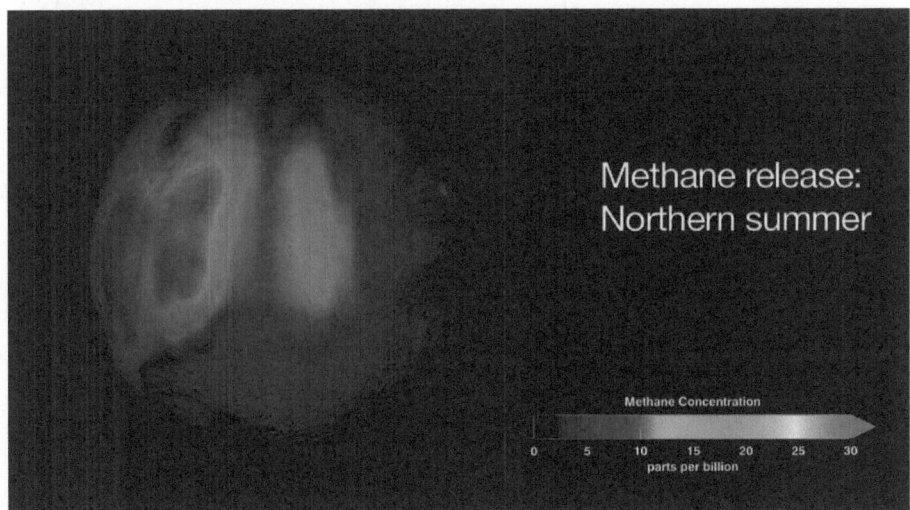

Figure 3.3 The distribution of methane on Mars.
(Courtesy of NASA.)

Figure 3.4 Io from the Galileo spacecraft.
Image of the pockmarked surface of Io. In the bottom left of the image you can
see a huge volcanic plume rising far above the surface. (Courtesy of NASA.)

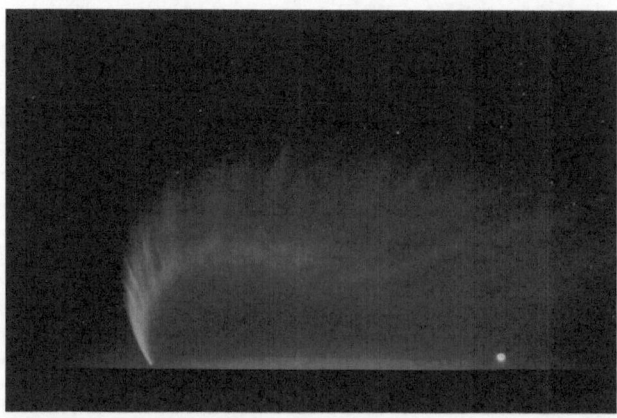

Figure 3.5 Comet McNaught and the Sun, setting over the Pacific.
(Courtesy of ESO; http://www.eso.org/public/.)

Jupiter is the giant of the Solar System, the largest planet by far. It is the first of the gas giant planets, the bulk of whose mass is made up of hydrogen and helium gas, unlike the rocky terrestrial planets of the inner Solar System. What one might think of as the surface of Jupiter is actually not solid at all. Instead, we are seeing the cloud tops of an atmosphere many thousands of kilometres deep. The structures in these cloud tops reveal a continuously churning, swirling, active atmosphere, with counter-rotating bands encircling the planet, and the famous great red spot, an eye-catching feature near the equator, which is a storm system bigger than the Earth that has lasted hundreds of years. You can get some idea of the structures in Jupiter's atmosphere from Figure 3.2, but movies show much more clearly what a dynamic system it really is[1].

Jupiter is also orbited by many moons, of which the four biggest, Callisto, Ganymede, Europa and Io, are very interesting in their own right. Originally discovered by Galileo, and thus known as the Galilean moons, they range in size from a diameter of 3100 to 5200 km. Io, the innermost, has a wonderful yellow colour as a result of the ejection of sulphur from the many volcanoes that cover its surface. Io is the most volcanic object in the Solar System since it is being continuously heated by the huge tidal effects of the immense gravitational field of nearby Jupiter.

Further out, Europa and Ganymede are the subject of much research and speculation since it is thought they may have vast oceans sealed beneath their icy outer surfaces. The NASA space probe Galileo spent nearly eight years making a detailed study of the Jovian system in the late 1990s and early 2000s, and future missions, such as ESA's JUICE (JUpiter ICy moons Explorer), will be looking at these moons in much more detail.

The next planet out from the Sun is Saturn, the second largest gas giant in the Solar System. The most famous aspect of Saturn is, of course, the ring system that surrounds it. The rings are made up of many small ice particles. These are likely the leftovers from a moon of Saturn that was shattered either by a collision or by tidal forces, but the age of the rings and the details of their formation process is still under discussion [27]. While Saturn's rings are certainly spectacular, the other three gas giants in the Solar System have ring systems of some sort, albeit much weaker than those of Saturn.

Saturn itself is in many ways just a somewhat smaller version of Jupiter, with similar internal structure and composition, though its atmosphere is somewhat less spectacular in colouration. As well as the rings, Saturn also has a plethora of moons, two of which are rather special. First there is Titan, which has a thick atmosphere consisting of nitrogen but with a small amount of methane. In the upper layers of the atmosphere the methane is broken down by ultraviolet light from the Sun, and the resulting hydrocarbon chemistry produces a layer of smog that shrouds the surface of Titan from view. This

---

[1]One such movie is available from the webpages associated with this book, but many others are also available online.

is a pity, since the surface of Titan is a very interesting place, with methane and hydrocarbon rain, producing rivers that flow into lakes and seas. Titan is actually bigger than the planet Mercury, but smaller than Jupiter's moon Ganymede. Secondly, there is Enceladus, the sixth largest of Saturn's moons. Until 2006 Enceladus was regarded as just another unexceptional moon of Saturn. That all changed when the Cassini spacecraft [47] discovered water geysers erupting from its southern hemisphere. The presence of these geysers implies the presence of a substantial body of liquid water beneath its icy surface, making Enceladus, a once disregarded moon, perhaps the most likely other place in the Solar System to host life as we know it.

The next two planets outward are also gas giants, but they are smaller than Jupiter and Saturn, and their composition is somewhat different, with less hydrogen and helium among their constituents. This has led them to be described as ice giants. Uranus is the seventh furthest planet from the Sun and the first of these ice giants. It has a radius four times that of Earth, and is largely made up of water, methane and ammonia. Methane clouds in its upper atmosphere give it an attractive blue colour (see Figure 3.2).

All the planets in the Solar System lie in the same orbital plane, known as the ecliptic, with their rotational axes at right angles to this plane. Uranus is odd in that its rotational axis nearly points along the ecliptic. This means that at some points in its year its north pole is always pointing roughly towards the Sun, and its south pole away from it, giving it seasons unlike any planet in the Solar System. This is probably the result of a catastrophic collision early in the planet's history. Uranus, like the other gas giants, has a plethora of moons, with 27 currently known.

Neptune is the last planet in the Solar System, the fourth gas giant, and the second ice giant. It too is made largely out of methane, water and ammonia, with a hydrogen and helium atmosphere tinted blue by methane. It has the usual banded cloud structures common to all gas giants, and giant storm systems not dissimilar from Jupiter's Great Red Spot. It has fourteen known moons, the largest of which, by far, is Triton. Among the large moons of the Solar System, Triton is unique in having a 'retrograde orbit', meaning that it orbits Neptune in the opposite direction to that in which Neptune rotates. This means that Triton could not have formed at the same time and place as Neptune. Instead, it must have once been a free-floating body that was captured by Neptune during earlier stages of the Solar System.

Unlike with all the other planets, dedicated space missions have not yet been sent to Uranus and Neptune, so our knowledge of these objects is not as well developed. Instead we have to rely on the results of fly-by observations by the Voyager 2 mission, and observations from telescopes on, or around, Earth. Beyond Neptune, we don't even have studies from the Voyager probes, so all we know about the outer reaches of the Solar System are based on Earthly observations and studies of objects which visit the inner system from time to time in the form of comets.

At this point we end our brief survey of the Solar System. There is a lot more to tell, including the demotion of Pluto from planet to minor planet. Much of this relates to observations made in the infrared, so we will cover it in more detail later in this chapter.

3.3   INFRARED ASTRONOMY IN THE SOLAR SYSTEM

All of the eight planets, their moons, the asteroids, and many of the other constituents of the Solar System, as we shall see, were discovered through optical observations. In the case of most of the planets these observations were conducted by the unaided human eye. Many of the planets and a number of other bodies have been visited by spacecraft that are able to make in-situ measurements that are very difficult to make from a distance. Given all this competition, you might expect that there isn't very much for infrared astronomy to work on in the Solar System, but that would be very wrong.

Infrared astronomy provides two unique capabilities. Firstly, by operating at wavelengths close to those where the black body emission of an object, or part of an object, peaks, it can be used to measure temperatures and temperature changes. This is especially important in the outer reaches of the Solar System where the light from the Sun is very weak, and so the objects themselves are quite cold. Secondly, many chemical compounds found on Solar System objects or in their atmospheres have characteristic spectral features at infrared wavelengths. Water ice, for example, which is nearly ubiquitous on moons and other Solar System bodies beyond Mars, has characteristic features at wavelengths of 3 and 15 microns, in the near and mid-infrared. The same goes for methane, carbon dioxide and many other chemical species. Some of the spacecraft sent into the Solar System carried infrared instrumentation, and made in situ observations at these wavelengths. Infrared observatories from the ground or Earth orbit can observe when these missions have expired, provide capabilities that were never available to the visiting spacecraft, and can look at the many Solar System objects that have yet to be visited by robotic probes.

3.4   PLANETS - PEERING THROUGH CLOUDS, STUDYING ATMOSPHERES

Many of the infrared observations of well studied planets, like Venus and Jupiter, are filling in gaps in the big picture results obtained by interplanetary missions. The surface of Venus, for example, which cannot be seen in the visible part of the spectrum thanks to the high altitude obscuring clouds of sulphur dioxide, has had its surface mapped in detail using radar techniques by the Magellan mission. Observations using interplanetary radar from the ground complement these results, as do observations in the infrared which can also see through the clouds. What all these studies reveal is a surface geology rather different from that of the Earth, with no signs of continental plates. But there are large plains of lava, giant shield volcanoes, and little evidence for impact craters, implying that the surface we are seeing is relatively young, about 800 million years old, while the Solar System itself is about 4.5 billion years old.

The picture that has emerged is of a planet similar to Earth in size and constituents, but that has a very different geology. One idea is that Venus very occasionally experiences a planet-wide volcanic event, with essentially the entire planet being resurfaced by new flows of molten lava [17].

One of the most important roles for infrared observations is probing the atmospheres of planets. This can be done by spacecraft or by ground-based observations. A good example of such studies are observations of methane in the atmosphere of Mars. Methane is an important molecule in the search for life. There is now plentiful evidence that liquid water once flowed on the surface of Mars. This means that conditions favourable for the development of life may have existed there, and life might have evolved. If it remains on Mars to this day, life is likely to be hiding below the surface, possibly in deep aquifers, or within the soil and rock of the planet. Direct observations of life hiding away in such places would be very difficult, since we would have to drill down into the soil and rock. So, instead of looking for life itself, we look for its waste products.

On Earth, one of the waste products of many anaerobic life forms - those that do not need oxygen to survive - is the gas methane. It's what gives rotting swamp water its distinctive smell. So a search for methane in the atmosphere of Mars might be a way to look for signs of life. However, methane is also produced by other processes, including geological activity, so the results need to be considered carefully.

Observers at the Infrared Telescope Facility (IRTF) and Keck telescopes in Hawaii observed Mars in the near-infrared, looking for spectral features from methane at wavelengths around 3 microns [115]. They mapped out the strength of this methane emission across the face of Mars, and monitored how this emission changed over several years. What they found is intriguing. It seems that methane emission from the surface of Mars is not evenly distributed (see Figure 3.3), but comes from specific locations. The emissions also peak in the northern hemisphere summer. A similar map of methane on Mars has also been produced using data from the orbiting Mars Global Surveyor (MGS) spacecraft [60].

It is tempting to interpret these results as coming from something alive that is thriving in the heat of the Martian summer, but the story gets more complicated. The ground-based observations were published in 2009 and based on data taken from 2003 to 2006, while the MGS results cover a period of about 10 years from 1996 to 2006. In 2011, the Curiosity rover landed on Mars, equipped with far better instruments than any previous rover, and able to look for methane in tiny quantities. While the rover isn't anywhere near the sites of the claimed methane emission, some of the released methane should have spread around the planet, allowing it to be detected by Curiosity's very sensitive instruments. But, in 2013, it was announced that Curiosity had detected no methane at all - not a single molecule.

We do not, at the time of writing, know what is going on here.

As with any scientific study, even those published in prestigious journals, there is room for reinterpretation and for unnoticed mistakes. The origin of the disagreement in this instance is unclear and there are good reasons to believe both sides of the story. The final conclusions are yet to be written.

3.5 MOONS: ICY, VOLCANIC, SMOGGY

If we are looking for life in the Solar System, then the moons of the giant planets are among the most interesting places. We suspect there are subsurface oceans of liquid water on Europa and Ganymede, in orbit around Jupiter, and on Titan and Enceladus in orbit around Saturn. In the case of Enceladus, there is direct evidence for liquid water from the water plumes found by the Cassini Spacecraft, and there is now evidence for similar plumes of water being ejected from Europa. Given that these plumes exist, how do we find out what is going on in the subsurface oceans that feed them?

One way of doing this is to build a new large spacecraft mission to go to the moon and make detailed studies. This can certainly lead to spectacular results, but it takes a very long time - a typical new large space mission will take at least ten to fifteen years from approval to launch and operation - and a lot of money. The alternative is to fall back on the observatories available on and around the Earth, and to make observations of these phenomena from a distance. These observations will also be helpful in working out where to send expensive future space missions.

In the case of the ice plumes, the most interesting things to look for are organic molecules. Methane, ammonia and carbon dioxide as well as water have been found in the plume by Cassini, along with a range of trace organic molecules, such as alkanes, alkenes, nitriles, aldehydes and alcohols [110]. One of the key questions is whether these molecules just reflect the original composition of Enceladus, and what has landed on its surface over time, or whether there are internal chemical and possibly biological processes that are producing these molecules today. For Enceladus, the Cassini spacecraft is still operating, so further in-situ studies can take place. In the case of the newly discovered plume on Europa, though, no spacecraft is currently operating around Jupiter. The next to arrive there will be JUICE, but it won't get there until 2030. In the meantime, observations from Earth will have to fill the gap.

A similar situation has happened in the past. In 1977, the Voyager spacecraft found that Io, the innermost moon of Jupiter, has tremendously powerful volcanoes (see Figure 3.4). The Voyagers were only fly-by missions, passing through the Jovian system on the way to the other outer planets. The Galileo spacecraft wouldn't reach Jupiter until the end of 1995, so anybody interested in Io's volcanoes had to rely on ground-based observations until then.

Fortunately, the revolution in infrared detectors that led to near-infrared camera chips was well underway in the 1980s, and new generations of infrared cameras, sensitive to the 1000-degree heat of Io's volcanoes, became available.

The IRTF has been running a monitoring campaign, imaging Io whenever it is visible, to study the range and variability of its volcanic activity. These observations continued to be useful even when Galileo had reached Jupiter, since the satellite wasn't able to observe Io all the time. More recently, larger telescopes, such as Keck, using systems to eliminate the image distortions produced by turbulence in the atmosphere, have produced even better images of volcanic eruptions on Io. The Hubble Space Telescope has also made its own contribution, and has even been able to measure the size of the volcanic ejection plume from an eruption from Pele, one of Io's most powerful volcanoes.

Like that of Venus, the surface of Saturn's moon Titan is obscured by clouds high in its atmosphere that make optical observations impossible. The clouds of Titan, though, are not sulphur dioxide but instead are a mixture of hydrocarbons not unlike smog. Work to use ground-based near-infrared observations to look through this smog started early, finding evidence for varying surface features across the moon [71]. When Cassini arrived at Saturn in 2005, this mapping work made rapid progress, with both infrared instruments and radar mappers producing our first full maps of the surface.

At the same time, aided by results from the Huygens lander that was launched from Cassini, the detailed structure and make-up of Titan's atmosphere was determined. What emerged was a world where methane, in many ways, takes the place of water, with methane lakes or seas towards the cooler poles, and methane rain replenishing them. Methane is gradually lost from this cycle as photochemical processes in the upper atmosphere turn methane into more complex hydrocarbons which form the smog, and eventually fall to the surface to form dunes in the equatorial regions [103]. If methane is gradually lost in this way, it needs to be replenished, otherwise, in the long term, over millions of years, it will run out.

More advanced ground-based observations in the near-infrared seem to have solved this problem. Observations from the Gemini North telescope have found that Titan's methane clouds seem to be produced above a small number of specific locations [143], a result that agrees with independent observations from Cassini [72]. This suggests that methane is being replenished by material inside Titan, and is emerging from specific sites of what is known as cryovolcanism, volcanic-like activity but at much lower temperatures where water ice, for example, plays the role of rock on Earth.

3.6 ASTEROIDS

While planets and their moons make easy targets for our space missions, the asteroids, lying in the asteroid belt, are a rather more difficult target. They are spread across the entire Solar System, and come in a wide range of different sizes and types. At some level, even finding some types of asteroids is difficult.

Asteroids come in several different classes, defined by what they are made of and by their surface characteristics. The most common asteroids are classified as C-type. They are dark in colour and are made out of carbonaceous material which contains a variety of compounds based on carbon. About 75% of asteroids are C-type.

Next most common are the S-type asteroids. These are lighter in colour than the C-type as they are largely made out of silicate material, containing a variety of compounds of silicon. You might think of these asteroids as being 'stoney' in nature. About 17% of asteroids are S-type.

Finally come the X-type asteroids. These don't fit into the C or S classifications and are largely made up of metals such as iron and nickel. These make up most of the remaining 8% of asteroids.

This is a somewhat simplified classification scheme, roughly equivalent to one published in 1975 [28] based on observations in the optical and near-infrared of 110 asteroids. There are several more complex and more recent classification schemes based on more comprehensive observations, but these basic groups remain.

The traditional way of finding asteroids is to take observations in the optical. Data are taken at two or more different times. The background stars in these images will remain fixed, but asteroids, which are orbiting the Sun, will move from one image to the next. The longer the time difference between the images, the further the asteroids will have moved. The light being seen from asteroids in these observations is light from the Sun reflected back to us from the asteroid's surface, since asteroids are nowhere near hot enough for them to be able to directly radiate at optical wavelengths. Typical temperatures for asteroids in the main belt vary from about 160 to 200K, depending on how far away they are from the Sun and how reflective their surface is.

The difficulty with a scheme that relies on light from the Sun being reflected by an asteroid is that many of them, especially the C-type carbonaceous asteroids, have a very dark surface. This means that little of the light that hits them is scattered away. Instead it is absorbed, and serves to heat the asteroid. C-type asteroids can reflect as little as 3% of the light that falls onto them. This can make them very faint in optical asteroid searches - you could think of them as stealth asteroids. The amount of light reflected also depends on the size of the asteroid - the bigger it is the more light it scatters - so small, C-type asteroids are the hardest to detect.

Operating in the mid-infrared provides a useful alternative way of detecting asteroids. The thermal emission of an asteroid at about 200 K will peak at a wavelength of about 15 microns, so observations with a sufficiently sensitive mid-infrared instrument should be able to pick up asteroids by their own thermal emission, independent of their reflectivity. The whole sky was searched for such objects by the IRAS and Akari satellites [165] [162], while smaller regions of the asteroid belt were studied to greater sensitivity by ISO [161] and Spitzer [111]. The general conclusion from this work is that optical searches miss a significant number of asteroids - the Spitzer study more than doubled the number seen in their target field - but the data match our current understanding of the formation and evolution of asteroids.

Figure 3.6 Vesta from the Dawn spacecraft.
(Courtesy of NASA.)

Very few asteroids have been studied in any significant detail. Asteroid Itokawa, a small (500 m in length) asteroid that has been nudged out of the asteroid belt and crosses the Earth's orbit, was studied by the Japanese Hayabusa mission in 2005. The spacecraft rendezvoused with the asteroid, took images and other data, and even sent a small lander down to the surface. The conclusion of this study was that Itokawa is not a solid body, but rather a 'rubble pile', a conglomeration of rocks of various sizes and shapes that have adhered together over time.

More recently the NASA Dawn mission was launched to make in situ observations of two of the larger asteroids in the asteroid belt, Vesta and Ceres. In July 2011, Dawn reached Vesta and spent over a year studying it. Vesta is the third largest asteroid, in terms of size, with an elliptical shape about 500 km across in two dimensions, and about 450 km in the other. The full results from Dawn's study of Vesta are not yet available, but it is already clear (see Figure 3.6) that it is a complex body, with an internal structure differentiated into a crust, mantle and core, in many ways similar to the Earth or Moon.

The largest asteroid in the asteroid belt is Ceres. It is roughly spherical in shape, with a radius of about 480 km. Ceres is the next target for the Dawn spacecraft, which is due to arrive there in April 2015. The nearly spherical shape suggests that Ceres is also a differentiated object, with a core, mantle and crust, like Vesta and larger bodies. Ceres is formally classified as a minor planet.

Even before Dawn arrives there, Ceres has been in the news. Observations from the Herschel Space Observatory have found water vapour being released from two distinct regions of its surface [94]. The source of this water is unclear. We could be seeing something similar to the release of water from Enceladus and Europa, implying the presence of a large subsurface volume of water, possibly kept liquid by the heat of radioactive decay. This would be an astounding discovery as it would mean there is another place in the Solar System where liquid water exists, and thus potentially life. Alternatively, the water may be evaporating off a layer of water ice lying just below the surface of Ceres, in a similar, albeit slower and much less spectacular, way to that in which water and other volatiles are released from a comet nucleus as it passes close to the Sun. Whichever of these possibilities is correct, Dawn's forthcoming encounter with Ceres has become even more exciting.

3.7 COMETS

Comets are among the most spectacular of astrophysical phenomena. Every few decades one comes along that is visible to the naked eye in daylight, even from the largest, most light-polluted cities (see Figure 3.5). In previous generations the appearance of a comet was seen as a bad omen that great changes or disasters were about to arrive. These days we know that this is not the case, unless we're very unlucky and the comet hits our planet. Instead, the arrival of a comet is a great opportunity for astronomers, since comets are messengers from the outer reaches of the Solar System, nudged towards us by some distant gravitational interaction, and carrying with them information about how the Solar System formed.

The most spectacular part of a comet is its tail, but this is just made of gas and dust released from the nucleus, which is where nearly all the mass of a comet lies. We currently think of a comet's nucleus as being something similar to a dirty snowball. It consists of a mixture of rock, dust and ices, including frozen water, and solid carbon dioxide, carbon monoxide, ammonia and methane. In the outer Solar System, all the volatiles remain frozen. When a comet approaches the inner Solar System everything begins to warm up, and the volatiles start to boil off, expanding away from the nucleus and carrying away some of the dust as well. The result is the tail or, to be more precise, the two tails, since the dust and gases, ionised by the Sun's ultraviolet emission, react differently to the solar wind. The dust and gas in the comet's tails, and in the atmosphere they form around the nucleus, known as the coma, are all prime targets for astronomers, including infrared astronomers who will be studying, among other things, the properties of the gas molecules released from the nucleus.

The dust released by a comet as it passes through the Solar System doesn't just disappear. Instead, a comet leaves behind a trail of dust in its orbit which, if it is a short period comet, gets regularly replenished. Some of these dust trails will cross the orbit of Earth, and when the Earth passes through one of them, we get meteor showers like the Leonids (mid-November, coming from comet 55P/Tempel-Tuttle) and the Perseids (mid-August, coming from comet 109P/Swift-Tuttle) as the dust particles burn up in our atmosphere.

Cometary dust, and dust expelled by asteroid collisions, produce a cloud of small particles in the inner Solar System. This dust reflects sunlight, producing a faint, diffuse glow in the sky, visible shortly after sunset or before dawn, known as the Zodiacal Light. Thermal emission from Zodiacal Light dust particles can be seen in infrared maps of the whole sky, such as those produced by the IRAS satellite (see Figure 6.4).

There are two types of comets, distinguished by their orbital periods. There are short period comets, like Comet Halley, that return to the Solar System every two hundred years or less. Comet Halley has an orbital period of 75 years and will next be seen by us in 2061. Short period comets have prograde orbits, meaning that they orbit the Sun in the same direction as the rest of the planets, and have low inclination angles, so they travel roughly in the plane of the ecliptic. Long period comets, with orbital periods greater than about 200 years, have very different orbital characteristics. Their orbits can be retrograde as well as prograde, and they can have a wide range of inclination angles.

Comets have relatively short lives since they will eventually run out of volatiles. Some comets have even shorter lives than that. Shoemaker-Levy 9 was captured and eventually collided with Jupiter, and comets are frequently seen hitting the Sun by satellites monitoring solar activity. So, since comets keep coming, there must be reservoirs of potential comets in the outer Solar System. In fact, since there are two different classes of comet, there have to be two different classes of reservoir.

The short period comets must come from a collection of bodies orbiting the Sun the same way, and in the same plane, as the planets. Their orbital period suggests that this reservoir should lie somewhat beyond the orbit of Neptune, which has an orbital period of 165 years. This idea was independently suggested by astronomers Gerald Kuiper and Kenneth Edgeworth, and it is now largely known as the Kuiper belt. As we shall see in the next section, Pluto was the first Kuiper belt object to be discovered, not that it was realised at the time, and we now know of many more.

The long period comets must come from somewhere else, somewhere further away and with no preferred orbital plane or direction. This idea was suggested by Ernst Öpick and, somewhat later, by Jan Oort, and is now known as the Oort Cloud. The existence and properties of the Oort Cloud remain hypothetical, as no objects in the Oort Cloud have ever been directly detected. Nevertheless, something like it must be there to explain the existence of long period comets. Models of the Oort Cloud suggest that it might extend out as far as several light years from the Sun. If that is the case, and if every solar system has a similar cloud, then the outer reaches of our system's Oort Cloud might overlap with that of our nearest neighbour star, Alpha Centauri.

3.8   THE OUTER SOLAR SYSTEM AND THE PROBLEM OF PLUTO

The first body beyond Neptune to be discovered was given the name Pluto and it was described, for a while, as the Solar System's ninth planet. Pluto is in fact very small for a planet, smaller than many of the larger moons, including our own. Nevertheless, in the absence of any other objects being discovered that far out, Pluto's status as a planet went unchallenged. All that started to change in 2005, with the discovery of Eris [26], an object almost the same size as Pluto, lying in an elliptical orbit, significantly inclined with respect to the ecliptic. The discovery of several other Pluto-sized outer Solar System objects was announced not long after, including Makemake and Haumea. These objects, along with Pluto, are all termed Trans-Neptunian Objects (TNOs), and are all associated, in some way, with the Kuiper belt.

As the list of large TNOs grew, the status of Pluto as a planet began to become a problem. There are now eight other TNOs, in addition to Pluto, with diameters greater than 1000km, with Eris being almost the same size as Pluto. To avoid an avalanche of new objects being classified as 'planets', the International Astronomical Union revised the definition of planet in August 2006. The new definition stated that a planet had to be in orbit around the Sun, had to be sufficiently massive to reach hydrostatic equilibrium, and thus a spherical shape, and had to have cleared its orbital neighbourhood of other debris. Objects which do not meet the third of these criteria are classified as minor planets.

The discovery of Kuiper belt Objects other than Pluto, thus led to Pluto's reclassification as a minor planet. This has proven controversial in some quarters, but the alternative would be to add many more names, such as Eris, Haumea and maybe even Ceres, to the list of planets in our Solar System. And that list would keep expanding, as future observations of the outer Solar System are likely to find more large bodies that could be classified as minor planets. I, for one, would rather limit the number of planet names to the eight I now know, rather than have to keep adding more, and, knowing astronomers, much more contrived names for the foreseeable future.

3.9   THE FORMATION AND EARLY EVOLUTION OF THE SOLAR SYSTEM

Our brief tour of the Solar System has now gone as far as it can, reaching into the Oort Cloud and, possibly, groping towards the Oort clouds of neighbouring star systems. From this distant perspective we can look back at our local solar system and get some ideas about its formation and early evolution.

The first person to think in detail about the formation of the Solar System was the 18th-century French scientist Pierre-Simon Laplace. The key observations that drove his analysis are that the planets all lie in a single plane, the ecliptic, and that the orbits of the planets and the rotation of the Sun are in the same direction. This suggests a strong linkage between the angular momentum of the Sun and the planets.

Laplace thought back to what might have come before the formation of the Sun and Solar System. He envisaged a vast cloud of gas and dust, sitting in space, changing purely as a result of its own gravitation. This pre-stellar cloud will, unless it is perfectly symmetrical, gradually collapse along the axis with the greatest density gradient, becoming flatter and gradually more disk-shaped. At the centre of this collapsing structure will be the body that eventually forms the Sun. This will gradually accrete more and more material onto itself, becoming denser and hotter. At the same time, the collapsing gas cloud will become an ever more flattened disk, the material of which will be orbiting around the young star at its core.

The material of the disk will be mostly hydrogen and helium, since these are the most common elements in the universe, with some heavier elements like carbon, silicon and oxygen mixed in. Many of these will be in the form of frozen volatiles, such as water, carbon dioxide or other ices. Silicon compounds, in contrast, along with a small fraction of metals such as iron and nickel, will be in a more heat resistant, refractory state, as small particles of rock or solid metal. These particles will gradually start to clump together within the disk and, after a while, these clumps will become large enough to attract more disk material onto themselves through gravity. They are becoming planetesimals, small planets in the process of formation.

The planetesimals will, at this point, be a mixture of rock, metal and volatiles. However, once the young star becomes hot and dense enough to start fusing hydrogen into helium, that will change. In the inner parts of the young solar system, the heat of the new star is sufficient to boil away volatiles. Water ice, and frozen gases like carbon dioxide, methane and ammonia will all be driven away from the inner parts of the forming solar system, preventing the planets closest to the star from becoming gas giants like Jupiter or Uranus.

Further away, though, the light of the young star will not be bright enough to boil away this volatile material. The point at which this process stops is known as the snow line, since beyond it water ice, and thus snow, is able to survive. In the early days of our own Solar System, the snow line lay some way beyond the orbit of Mars. This is why Mars and the inner planets are rocky terrestrial planets, while the outer system is dominated by gas giants.

The early days of the formation of the Solar System end with eight planets orbiting the Sun where the protoplanetary disk once stood, with all of the material of that disk collected by the planets themselves. Those planets inside the snow line have lost all their volatiles, while beyond the snow line we have planets that have collected vast amounts of hydrogen, helium and other volatiles around the rocky seeds at their cores. Beyond these eight planets lies the debris of planetary system formation, in the form of the Kuiper belt, made up of the parts of the protoplanetary disk too low in density to be able to form a planet of their own.

This leaves us with a problem. In this scenario the inner planets, including the Earth, would have no atmospheres since they lost all their volatiles when the young Sun first ignited. Fortunately for us, the Solar System has a few tricks left. During the first 500 million years of the Solar System, it now seems that the gas giant planets, Jupiter, Saturn, Uranus and Neptune, moved about a bit in their orbits. Jupiter moved inwards a little, and the other gas giants moved outwards. The effect of this was to destabilise the young Kuiper belt. This resulted in the ejection of some of these volatile-rich bodies into the Oort cloud, while many others were diverted into the inner Solar System, producing a hail of volatile-rich comets and asteroids, many of which collided with the inner terrestrial planets. This all happened between 4.1 and 3.8 billion years ago, during a period known as the late heavy bombardment. The direct results of this bombardment can still be seen in the cratering record of the Moon, Mars and Mercury, but an even more important result of this phase in the early life of the Solar System was the arrival of liquid water on Earth, along with the other volatiles that form our atmosphere.

3.10   THE DISCOVERY OF EXOPLANETS

The planets of our own Solar System are no longer the only ones we know. Since the mid-1990s, when the first exoplanets were detected, an active industry has developed for detecting and classifying planets around other stars. At the time of writing nearly two thousand planets have been identified orbiting other stars, many of which are in systems containing multiple planets.

The two most successful methods of detecting exoplanets so far both involve looking at the parent star rather than the planet itself. The first of these, known as the radial velocity method, relies on the fact that planets do not actually orbit about the centre of their parent star, but the star and planet in fact both orbit about their mutual centre of gravity. Since stars are much more massive than planets, this centre of gravity will be very close to the centre of the star, but offset from it a little bit, the exact amount depending on the mass of the planet. This means that the star will move during the course of the planet's orbit. The amount that the star moves from side to side on the sky is very small, in fact too small for us to measure.

However, as well as this side to side motion, the star will also move towards and away from us. Using precisely calibrated high resolution spectra, the velocity with which the surface of a star is moving towards or away from us can be measured to accuracies of one metre per second or better. Massive planets, similar in mass to Jupiter or Saturn, can produce velocity shifts of as much as 12 m/s, which today are easily detectable. The first extrasolar planet to be detected around a Sun-like star was found using this method [108].

The second most popular approach is to monitor the light of a star and look for the small dip in brightness that occurs when a planet passes in front

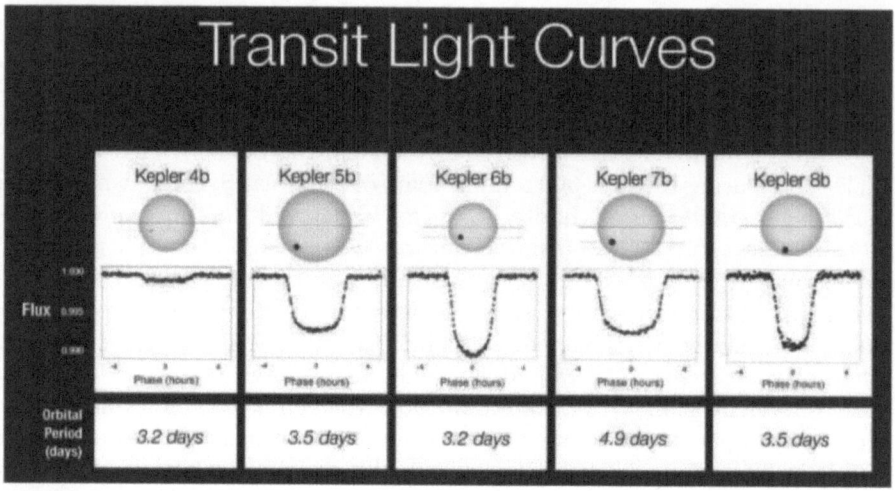

Figure 3.7 Light curves and diagrams showing the transits of the first five Kepler planet detections.
Courtesy of NASA [20].

of it. This is known as the transit method, as the passage of a planet in front of a star is called a transit[2]. The fractional dip in the light output of a star like the Sun amounts to about 1% when a planet the size of Jupiter passes in front of it, and about 0.01% for a planet the size of the Earth. Ground-based observations are capable of finding the dips coming from Jupiter-size planets transiting other stars [35], but the discovery of smaller planets requires more accurate measurements which are only possible in space, beyond the turbulent interference of the Earth's atmosphere.

Of course not all stars are aligned in such a way that their planets pass across them from our point of view. The chances of this depend on the size of the planet and how far away it is from its parent star. The planets in our own Solar System typically produce transits for only about 0.1% of potential extraterrestrial observers. Fortunately there are plenty of stars in the sky, so planet searches based on the transit method monitor the light output of many stars simultaneously. The Kepler mission, which has been one of the most successful planet-hunting projects to date, monitored the light output of 145000 stars continuously for three and a half years until a technical failure brought its mission to an end. Figure 3.7 shows some example planets detected by Kepler and the variation of light received from the parent star produced by the transits of these planets [20]. Such plots are known as light curves.

---

[2]Planetary transits also occur in our own Solar System. In 2012, for example, there was much excitement as it was possible to see the planet Venus pass in front of the Sun. The next time this will be visible from Earth will be in December 2117.

We now know of nearly two thousand exoplanets, with nearly five hundred multiplanet systems detected[3]. The holy grail of exoplanet studies is to find an Earthlike planet orbiting its star at a distance appropriate for liquid water to exist on its surface. Orbits where liquid water can exist are described as lying in the 'habitable zone', since we believe that liquid water is essential for life to arise. Unfortunately, our current detection methods are only just becoming good enough to be able to detect such planets, since they will produce only small radial velocity changes or small transit dips in their parent star's light curve. Nevertheless, there are a few candidates, and an analysis of all the confirmed and candidate planets currently in the Kepler database [128] suggests that 22% of Sun-like stars will have an Earth-sized planet lying in their habitable zone. This would imply that the nearest habitable planet might be just 12 light years away from us.

Exoplanet searches produced a number of surprises right from the start. The first exoplanets found were all high mass gas giants, comparable in size to Jupiter if not bigger, orbiting very close to their stars. This is not what you would expect given what we know, or thought we knew, about planet formation. Gas giants form well away from their parent stars, beyond the snow line, so that the young star cannot boil away the volatiles that lead to their high mass. And yet a large number of planets with masses equal to or greater than that of Jupiter, have been found so close to their star that their surface temperatures are over 1500 K. Does this mean that our entire picture of planet formation is wrong?

Fortunately, we do not have to go back to the drawing board. The explanation is that these 'hot Jupiters' have shifted orbit after their formation. They formed beyond their system's snow line, and then, as a result of interactions with other young planets and protoplanetary material, they moved inwards, towards their star. Hot Jupiters were very common among the early exoplanet searches not because they are that common in other solar systems but because they are by far the easiest type of planet to detect.

3.11 CHARACTERISING EXOPLANETS

Now that we are detecting an ever-increasing number of planets around other stars, the next goal is to start to characterise them, and to determine the constituents of their atmospheres. The first exoplanet atmosphere ever detected came as a result of careful observations of the dip in the stellar light curve at the start and end of a planetary transit, and seeing how this varies with wavelength.

For example, an atmosphere of hydrogen lying above the surface of a planet will produce a larger and longer transit dip at wavelengths corresponding to

---

[3]More planets are being announced all the time. The figures here are correct at the time of writing, but if you want the absolutely latest figures, and to explore the properties of all the known exoplanets yourself, you can consult websites like exoplanet.eu which collect all of the latest information.

hydrogen absorption lines than the transit dip associated with the opaque body of the planet itself. The same applies to other possible atmospheric constituents, and there are now several clear detections of this effect e.g. [30] and [168]. Indeed the latter study found hydrogen absorption so far away from the associated planet that it is likely that the atmosphere of this object is being gradually boiled away by the heat of its parent star. This same technique, of monitoring the depth of the transit as a function of wavelength, has been applied in the infrared to detect methane, water vapour and carbon monoxide in the atmospheres of hot Jupiters.

Light coming from the planet itself can also be detected by looking for what is called the secondary eclipse. This is where the planet goes behind the parent star. Since our telescopes do not provide sharp enough images to separate the light of the star from the light of the planet, when we observe these transiting systems we are always looking at the combined light of both objects. The contribution from the planet might be light from the star reflected by the planet, or it might be thermal emission from the planet's own heat depending on the details of the system or the wavelength of observation. The only times when the light received from the system is not the sum of emission from both planet and star are when the planet is transiting in front of the star, which produces the transit dip since the planet is absorbing some of the star's light, and when the planet goes behind the star, when the star obscures all the light we receive from the planet. The latter is known as a secondary eclipse. Accurate measurements of these secondary eclipses have found small dips in the light received in the infrared, allowing thermal emission from planets to be measured. Results from these studies have allowed the temperatures of some exoplanet atmospheres to be measured. Since these are hot Jupiters the temperatures are about 1000K. Combining these temperatures with the results from spectroscopy of the planetary transits themselves allows us to get a fairly detailed idea of the structure, constituents and behaviour of the atmospheres of planets many light years away from us.

All of these studies rely on having a planet fortuitously aligned with our line of sight so that we can see a transit. Things would be much easier if we could actually see the planets directly as they orbit about their star. This is very difficult since stars are much, much brighter than the planets that orbit them. However, technology is being developed that will be able to remove the light of the star and allow orbiting planets to be studied directly. This technology includes devices known as nulling interferometers and vortex coronagraphs. The eventual hope for these instruments is that they will be able to extract spectra for Earth-like planets orbiting other stars in their habitable zones, so as to search for any evidence for life on those planets. The signature of such life would be evidence of oxygen in the planet's atmosphere and, more specifically, the detection of ozone, produced from that oxygen, in the upper parts of a planet's atmosphere. Ozone has a very distinctive, deep absorption band at wavelengths around 10 microns in the mid-infrared. Comparison of mid-infrared spectra of the Earth, Mars and Venus demonstrate that this ozone feature is unique to the Earth in our own Solar System. The discovery of a similar feature in the atmosphere of an Earth-like exoplanet would be a clear indication that life exists elsewhere in the universe.

3.12 CONCLUSIONS

The study of our own Solar System has turned vague points of light that move around the sky into places that now have their own landscapes, geography and individualities. This was achieved using a combination of robotic probes and observatories based on the ground and in Earth orbit. The next decades will see a similar rapid proliferation of knowledge about planets and planetary systems around other stars. Our robotic probes won't be able to travel to these new places, but observing techniques are already being developed, as you have seen, which allow us to study the atmospheres and surfaces of these distant places. We may never be able to look at small pebbles on the surface of an earthlike planet orbiting a distant star, but there is a realistic possibility that we will be able to say something about its climate, the constituents of its atmosphere and some of the large scale features of its surface geography. The techniques to do these things are already being applied to the easier-to-observe giant planets.

While the quest for life elsewhere, whether beneath the icy moons of our own Solar System or on an earthlike planet orbiting a distant star, is a central part of this work, the study of planetary systems also tells us about the end point of the processes behind star formation. Understanding how stars form is another of the key projects of astrophysics. Infrared observations turn out to be very important in the study of star formation, as we will see in Chapter 5. However, before we can look at star formation in detail, we must look at the other product of the star formation process - the stars themselves.

CHAPTER 4

The Heat of the Stars

4.1   THE STAR-FILLED SKIES

The work of an observational astrophysicist takes you from observatory to observatory, so you might expect us to get plentiful opportunities to look at the sky. But that could hardly be further from the truth.

When observing, we spend most of our time in a telescope control room, monitoring our observations, and making sure everything is going right. More than that, the kind of things being observed by our telescopes are usually not the sort of thing that the eye can easily see - the fields are smaller, the depth greater - so even when we're in the control room we don't get a good impression of what the sky actually looks like. When a typical observing run is over we don't linger. We have things to do back at our home institutions, students waiting for us, and a stack of jobs that have been piling up while we've been away, And when we get home, since most universities are in major cities, plagued by light pollution, we still don't get to look up and see the sky.

Amateur astronomers, who have the time and leisure to look at what they want for as long as they want, usually have a much better appreciation of the night sky.

But tonight is different.

I'm visiting an observatory I've never used before in Spain, so I left myself an extra night at the telescope to make sure I could get my data properly transferred onto tapes in case there were any delays. My fears were proved right, and some extra time at the observatory was necessary, but now the copying is done, my data tapes are safely in my bag, and the night is still young.

It's summer. The sky is clear, the weather warm. My four nights of observations, studying colliding galaxies in the near-infrared, went very well, but there is nothing else to do tonight but sleep, or look at the sky.

So I hop on one of the observatory's bicycles, and head towards the telescope, then turn off the road, and find a comfortable place in the grass surrounding the buildings to lie down and stare upwards.

The sun has just set, so at first I can only see the brightest stars and planets. Conditions are good, so the stars aren't twinkling that much. Off to the left, a bright object drifts across the sky. It must be a satellite, something in low Earth orbit, to be moving so quickly against the background of the 'fixed' stars. A few moments later I notice that another bright, but slightly fainter, object is following the first, chasing it across the sky. I realise that the bright object must be the Russian Mir space station, and that the other is the Soyuz capsule, launched a few days ago, taking up supplies and a new crew.

The sky gets darker, my eyes get more dark-adapted and the stars get more numerous.

My view must be along the plane of our own spiral galaxy, since there seem to be so very many stars. They fill my field of view, seeming to almost overlap, and turning parts of the night sky from the inky blackness of a good astronomical site to a dull glow of blurred-together stars[1].

The stars I'm used to seeing from the city, against the glow of the light-polluted sky, always appear as points of undistinguished light, points of greyish luminosity amid the muddy dark. On a good night at home, I might just get an impression of reddish colour from the planet Mars. Colour is something that is banished from the sky in cities.

Not so here. For the first time since I centred a finding 'scope on a bright star to set a telescope's pointing, and saw that the star was a bright actinic blue, I can see the colours of stars. And they vary so much.

There are blues, not far off what I saw through that finding telescope a few years before, but there are also yellows and reds, all of different hues and brightnesses.

I already know that these colours are much more distinct when you see them with a telescope. The binoculars I'm carrying tonight make a huge difference, but a full-scale telescope can bring those colours to life for even the faintest star perceivable to the naked eye.

Hidden among the wonder of this vision of the sky are fundamental astronomical observations. The number of stars, and how their number changes with brightness, allowed William and Caroline Herschel to produce the first maps of the structure of our home galaxy, the Milky Way. But it is the colours of stars, and how that colour depends on their other properties, that tell the story of the life and death of stars.

4.2   THE MAIN SEQUENCE

The nature of stars has been a subject of speculation ever since we were human, and perhaps before, but it was only with the advent of astronomy as a science that we started groping towards some understanding of them, rather then telling comforting stories about the shapes they seemed to make in the

---

[1]My eyes aren't actually that good, so this effect was probably because of somewhat blurred vision. However, this kind of effect, where sources on the sky are so numerous that they blur together, is an actual astronomical effect known as confusion.

sky. The first step often taken when faced with a class of objects that we don't understand is taxonomy - trying to divide the objects up into different types. The hope is that an examination of what distinguishes the different types will provide the first clues to understanding the class as a whole.

The taxonomy of stars began in earnest in the early 20th century, and is based on both the visual colour of stars, discernible for the brightest stars by the human eye on a good, dark night, and by measurements of their optical spectra. The colours and spectra of stars led to their classification by spectral types which correspond to their surface temperatures. The spectral types devised in the 1920s by Annie Jump Cannon at Harvard College Observatory, as part of her work on the Henry Draper Catalogue of 400000 stars, are still used today. The types are, in order running downwards in temperature, O, B, A, F, G, K and M, with corresponding temperatures ranging from 40000 K to 2400 K. Our own star, the Sun, is a G2 star in this classification, with a surface temperature of about 6000 K.

The next step in understanding the stars was to determine their luminosity - the amount of power they are producing. This was achieved by studying how stars, or groups of stars, appeared to shift position on the sky with time as a result of parallax as the Earth orbits the Sun, or as the Sun moves through the galaxy. The range in stellar luminosities found from this analysis is very large - from less than 1/10000th of the luminosity of the Sun to more than a million times brighter. Most stars, as we shall see later, are smaller than the Sun and less luminous, but the stars you can see with the naked eye are generally more luminous than the Sun, since they can be seen by the eye from far greater distances.

The key result came when these various classifications and calculations were put together, and astronomers looked at how the temperature and luminosity of stars were related. This was done independently by two different people, Ejnar Hertzsprung in Europe, and Henry Russell in the US. The result of their work is thus known today as the Herzsprung-Russell Diagram, or H-R Diagram for short. This diagram can be seen in Figure 4.1 in a descriptive form, while an actual observational H-R diagram can be found in Figure 4.2.

Two things are striking about the H-R diagram. Firstly, there is a long, coherent track across the middle of the diagram. This is called the main sequence of stars. Secondly, some classes of star lie well away from the main sequence and must thus, in some way, be different from the normal, main sequence stars.

The fact that we see so many stars on the main sequence suggests that this is where stars spend most of their lives, and this turns out to be the case. Stars, once formed, only move off this line when they are going through some significant changes late in their life, or when they have already, essentially, died. We will deal with the late stages of stellar evolution, and what infrared astronomy has told us about this, later in this chapter. For now let us concentrate on the main sequence, on what it tells us about stars, and how infrared observations have expanded our picture of the normal life of stars.

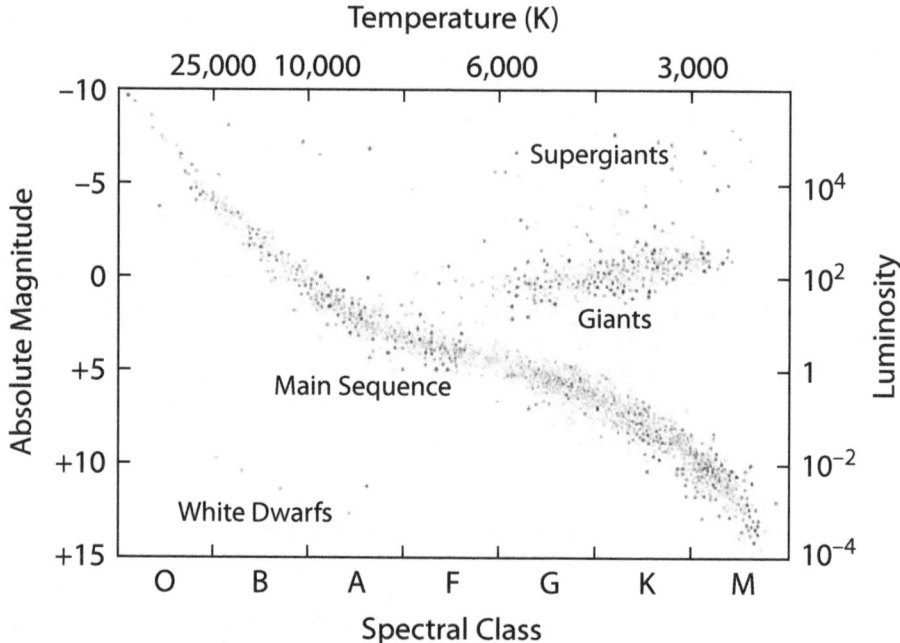

Figure 4.1 The Hertzprung-Russell Diagram.
A version of the Hertzprung-Russell Diagram, with spectral type on
the bottom and absolute magnitude, a measurement of luminosity, on
the vertical axis, showing different classes of star. (Courtesy of ESO
http://www.eso.org/public/.)

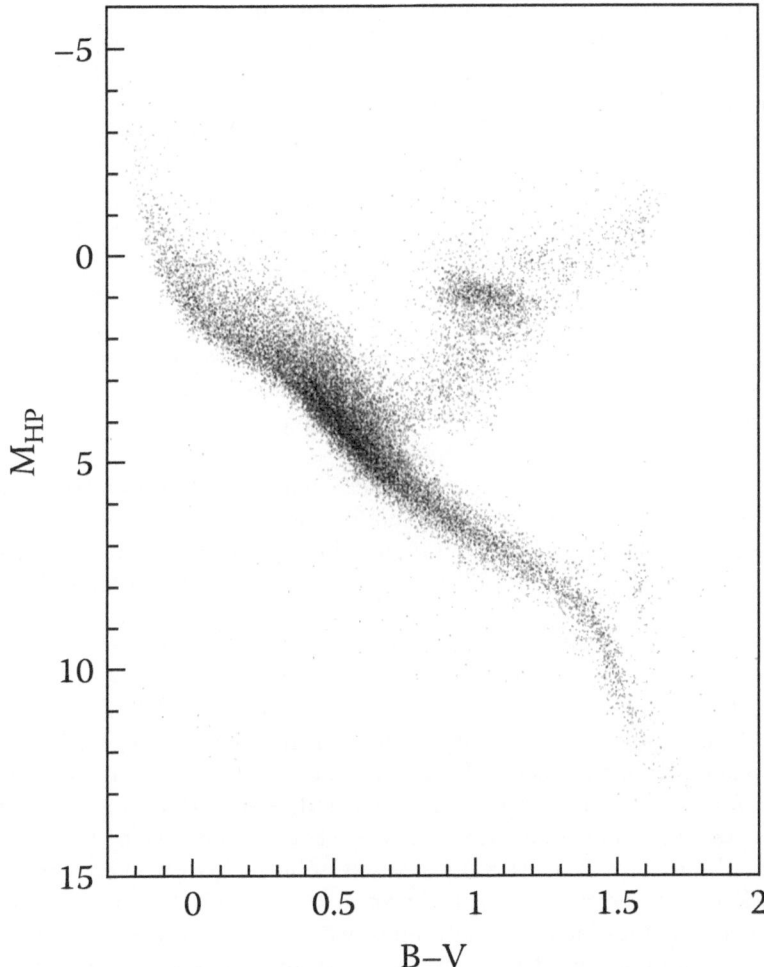

Figure 4.2 An observational Hertzprung-Russell Diagram.
Actual observations of the H-R diagram taken from the ESA Hipparcos mission. Hipparcos made accurate measurements of the properties of 118000
stars. For this version of the HR diagram a straightforward measurement
of colour, the difference in the star's brightness in two optical filters, blue
(B) and green (V), is used instead of spectral type. (Courtesy of ESA;
http://www.esa.int/ESA.)

4.3   THE LIFE OF STARS

Stars are massive balls of gas, kept hot and, as a result, kept from collapsing under their own gravity, by the heat generated inside them. The power source inside stars is nuclear fusion, converting hydrogen into helium. A star stays on the main sequence as long as the fusion of hydrogen into helium is its main energy source. Once the hydrogen runs out, and the star begins to burn heavier elements, its behaviour will change and it will leave the main sequence.

Many of the details of stellar behaviour on the main sequence were worked out before the fusion reactions that power them were understood. This is because, even in the early 20th century when scientists like Arthur Eddington worked out the basic rules of stellar structure, we knew how gases respond to different temperatures and pressures. With this understanding, and assuming an unknown central power source, Eddington was able to work out how a stellar atmosphere would behave. For example, using the simple assumption that the energy available to a star is proportional to its mass (which turns out to be correct) we can find that the lifetime of a star on the main sequence goes as the inverse of its mass cubed.

Our Sun, for example, is expected to remain on the main sequence for a total of about 12 billion years - it's currently about 4.5 billion years old, so is only now beginning to think about middle age [150]. A star three times the mass of the Sun, though, will stay on the main sequence for only ∼1/27th of that time - just 400 million years. Any heavy cousins of our Sun, formed at the same time, will have left the main sequence long ago. But there are compensations for this shorter life. It is said that 'a star that burns twice as bright burns half as long,' but, in this case, our higher mass star will burn very brightly, with a power output eighty times that of the Sun.

The details of the nuclear fusion reactions that power main sequence stars were worked out much later, once we understood that nuclear fusion can release large amounts of energy. There are two main routes by which hydrogen is turned into helium in stars. Our own Sun burns hydrogen into helium using what is called the p-p, or proton-proton chain. This is a series of reactions by which a helium nucleus, containing two protons and two neutrons, is built from the combination of four protons, and involves a variety of intermediate steps.

Higher mass stars are able to use the nuclei of the heavier elements carbon, nitrogen and oxygen (C, N and O) to catalyse fusion reactions, thanks to the higher densities and temperatures in their cores. The resulting process, called the CNO cycle, produces more energy than the p-p chain when it can operate. This gives rise to the kink in the main sequence for stars hotter than spectral class A. More details of the p-p chain and the CNO cycle can be found in [68] and other textbooks on stellar structure.

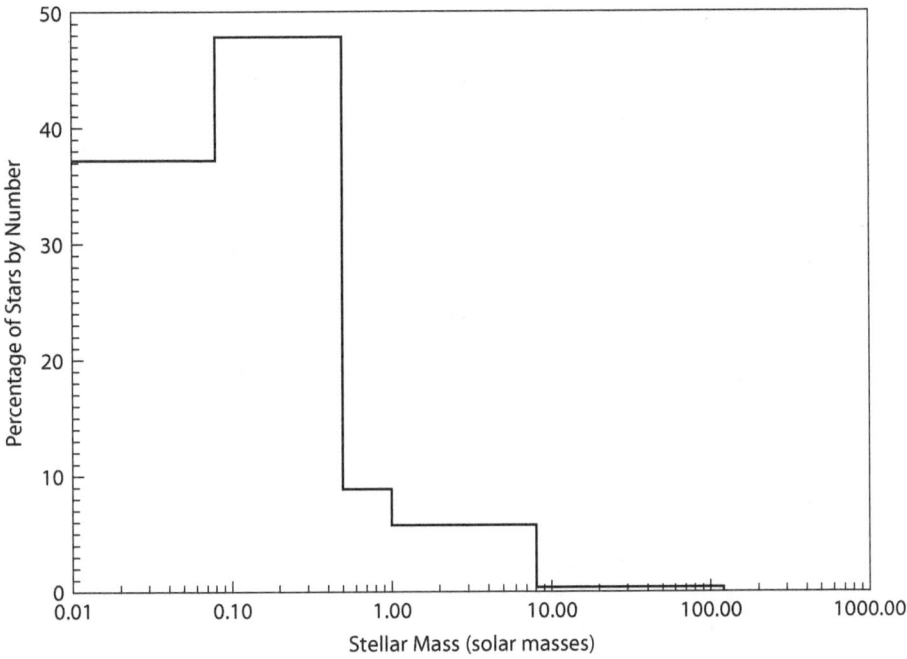

Figure 4.3 The stellar initial mass function.
The percentage of stars in our galaxy as a function of their mass, based on
the formulation by Kroupa [92]. As you can see, most stars are in fact lower
mass than our own Sun, and there are very few high mass stars.

4.4 THE SIZES OF STARS

The mass distribution of stars - what fraction of stars are low mass, what
fraction are high mass - is a central question in astrophysics. The stellar initial
mass function (IMF), as it is called, is important for understanding where
stars sit on the HR diagram. It is also related to the processes that form stars,
since these are what produce the IMF in the first place, and is important for
understanding the formation and evolution of galaxies, since it is largely by
the light of their stars that galaxies are studied. We will look at the origins of
the IMF and the story of star formation in Chapter 5, the nature of galaxies in
Chapter 6, and their formation in Chapter 7. The IMF touches on all of these,
which gives you some idea of how important it is. At the moment, though, we
will consider the determination of the IMF from an observational perspective,
and look at how observations of stars in the neighbourhood of the Sun lead
to plots like the one in Figure 4.3 that show how common are stars of various
different masses.

In concept, the IMF should be simple to calculate. You pick a representative volume of the local galaxy, and then count how many stars as a function of mass you find inside that volume. Unfortunately, what is easy conceptually is often very hard in practice. Direct measurement of the mass of a star, for example, is not very precise but, if you have enough stars, the overall trends can be determined fairly accurately. Getting enough stars in a well-defined volume, though, is a more serious problem.

Bright, high mass stars can be detected from a long way away, but the highest mass stars have very short lives - only a few million years for masses greater than 30 times that of the Sun. So finding stars of the highest mass, necessary for measuring how common they are, relies on catching them while they are still around. The best places to do this are where they form, since they don't last long, and won't have much time to move away from their place of origin.

Star formation, as we shall see in Chapter 5, occurs in giant molecular clouds (GMCs), regions of the galaxy where interstellar space is rich in hydrogen molecules, other molecular species, like carbon monoxide and water, and the dust grains that accompany them. The dust in these regions is a good absorber of light from any stars that form inside them. The energy absorbed by the dust grains heats them up, and they radiate it away in much the same way as a simple black body. The typical temperature of dust in these regions is around 30 - 50 K, so the peak wavelengths for this emission lie in the far-infrared. Giant molecular clouds are thus excellent targets for far-infrared telescopes like the Herschel Space Observatory.

Figure 4.4 shows the Herschel image of the W3 giant molecular cloud. This lies 6200 light years away from us and is the home for a substantial amount of massive star formation. High mass stars put out much more light than low mass stars, so they heat the dust near to them much more than their lower mass cousins. The blue regions in Figure 4.4 are the locations of the hottest dust, and thus the highest mass stars, more than ten times the mass of the Sun. Redder point sources, embedded in the larger scale filaments of dust and gas in this GMC, are the locations of lower mass stars and protostars. Studies of regions like W3 tell us about the relative frequency of high and low mass stars in their regions of formation. Also, as we shall see in Chapter 5, such studies reveal details of the process of star formation itself.

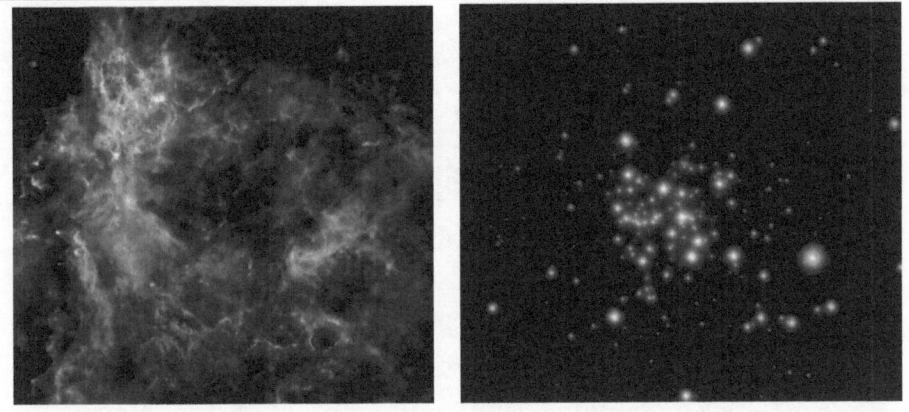

Figure 4.4 Left: Giant molecular cloud W3 from Herschel. Right: The Arches cluster in the near-infrared.

Left: The dust in W3 is shown as three colours: 70 microns in blue, 160 microns in green, and 250 microns in red. It is in the blue, hot dust regions that the highest mass stars have formed. (Courtesy of ESA http://www.esa.int/ESA.) Right: Near-infrared (J, H & K band) image of the Arches cluster near the centre of our own galaxy. (Courtesy of ESO http://www.eso.org/public/.)

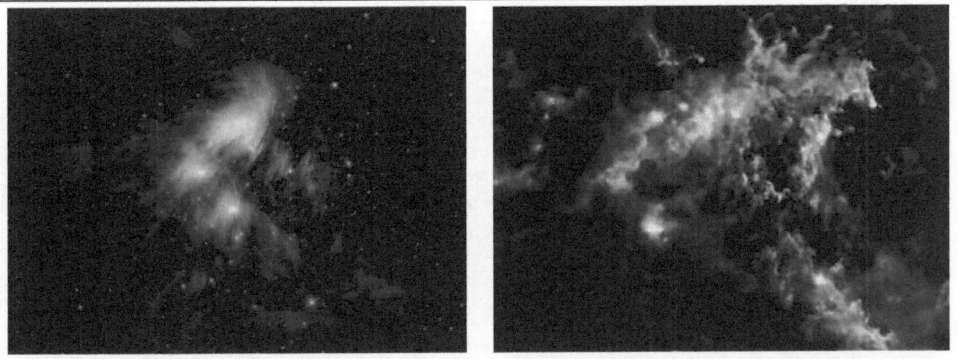

Figure 4.5 Left: The Pleiades star cluster in the infrared. Right: The Rosette Nebula from Herschel.

Left: WISE image of the Pleiades. Blue and cyan show the light from stars, green and red show dust emission. The dust is not associated with the star cluster but is just a cloud of interstellar material the cluster is passing through, being heated by its stars. (Courtesy of NASA). Right: Dust in the Rosette Nebula from Herschel. Globulettes that can form free-floating planets have been found here. (Courtesy of ESA http://www.esa.int/ESA.)

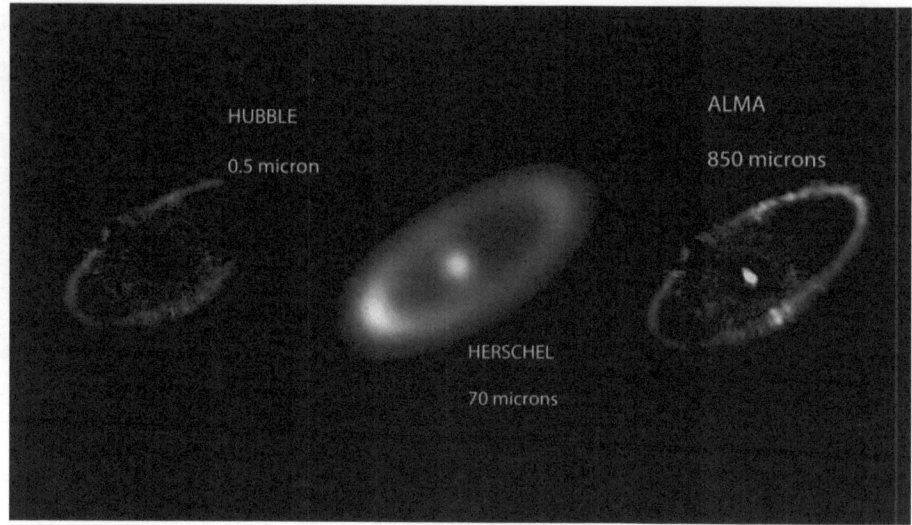

Figure 4.6 The Fomalhaut debris disk.
Hubble, Herschel and ALMA images of the Fomalhaut debris disk. ALMA observations are shown in red over Hubble observations in blue. ALMA only observed half the ring. (Courtesy of NASA, ESA; http://www.esa.int/ESA, and ALMA http://www.almaobservatory.org/.)

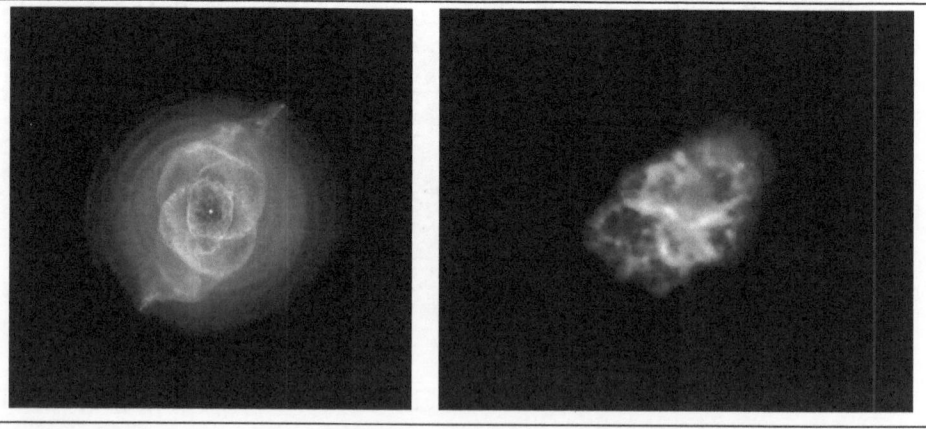

Figure 4.7 Left: The Cat's Eye Nebula. Right: Dust in a supernova remnant.
Left: The Cat's Eye, a planetary nebula, from HST. (Courtesy of NASA.) Right: The Crab Nebula from Herschel. 100 microns is shown as blue, 160 green and 250 red. The material seen here is dust produced during the supernova explosion. (Courtesy of ESA http://www.esa.int/ESA.)

Finding massive stars is not enough. They are rare, so when we find them they are going to be far away. This means it can be difficult to tell whether they are single massive stars or a binary pair of less massive objects. Even when you can find very massive stars, going from the number found to a number per unit volume can be difficult. For this reason, observations of discrete star clusters - groups of stars that formed at the same time and can be gravitationally bound to each other - are very useful, since the volume of the cluster can be measured and so the number of stars per unit mass can be determined more easily. The two best-studied large star clusters for this purpose are R136 (this lies in the larger 30 Doradus region, also known as the Tarantula Nebula, of the Large Magellanic Cloud (LMC), a satellite galaxy of the Milky Way), and the Arches cluster, which lies near the centre of our own Galaxy. In both cases near-infrared observations were necessary to fully understand these clusters, since longer wavelengths were needed to see through the dust in the host star forming region and reveal the stars inside. In the case of the Arches cluster, the near-infrared was also needed to see through the dust that lies between us and the centre of our own galaxy. You can see near-infrared images of these clusters, taken in the J (1.1 micron) H (1.6 micron) and K (2.2 micron) bands in Figures 4.4 and 4.8.

It is generally thought that the highest mass star that can be produced by normal star formation processes is 150 times the mass of the Sun [93]. This is what the observations of the Arches cluster find, for example. The observations of R136 shown in Figure 4.8, however, suggest the presence of stars with masses as high as 300 times that of the Sun [39]. This apparent contradiction was recently solved by the realisation that massive stars in dense star clusters might merge together to form even more massive stars [11].

At the other extreme of the IMF, we are faced with a rather different problem. High mass stars might be rare, but they are very bright and thus can be detected from a long way away (170,000 light years in the case of R136, for example). Low mass stars, of spectral class M, in contrast, are very common - they make up about three quarters of main sequence stars in the neighbourhood of the Sun. They are cool, with surface temperatures less than half that of the Sun, and small, with masses ranging from 60% to 10% that of the Sun. They are also very faint, with total power outputs as low as 0.02% that of the Sun.

Going from the number of low mass stars in the solar neighbourhood to a determination of the low mass IMF is complicated by the fact that our best map of the local stars so far, from the Hipparcos satellite, is not very sensitive, so we only have a proper census of faint stars within about 15 light years. For this reason, we once again have to resort to more sensitive observations in either star clusters, or along specific directions in the sky. The Pleiades, a nearby cluster of stars visible to the naked eye, is a favourite target for such searches. Recent near- and mid-infrared observations of this cluster [158] have allowed stars with masses down to 10% of the mass of the Sun to be detected, improving our knowledge of the low mass end of the IMF. More recent data on

Figure 4.8 R136, a high mass star formation region in another galaxy. Near-infrared image of the R136 cluster in the Large Magellanic Cloud from the Hubble Space Telescope. (Courtesy of NASA.)

the Pleiades and other star clusters from the WISE satellite, which surveyed the entire sky at four near and mid-infrared wavelengths (3.4, 4.6, 12 and 22 microns) will improve things still further. WISE's image of the Pleiades is shown in Figure 4.5.

4.5   BEYOND THE BOTTOM OF THE MAIN SEQUENCE

So far we've been discussing the properties of stars without having a precise definition of what kind of object is classified as a star. The key characteristic of a main sequence star is its source of energy generation - the fusion of hydrogen into helium. As we look at lighter and lighter stars at the bottom of the main sequence, we reach a point at which the densities and temperatures at the cores of these objects are insufficient to trigger sustained hydrogen fusion. This happens at masses less than 0.08 times the mass of the Sun, equivalent to about 80 times the mass of the planet Jupiter. At this point we start calling the objects 'brown dwarfs'.

These objects, however, are not without a means of generating energy. Normal hydrogen is made up of a single proton being orbited by an electron. As discussed in Chapter 1, a heavier form of the same element, called deuterium, exists, which has a nucleus made up of a proton and a neutron, once again orbited by an electron. There is roughly one deuterium atom for every six thousand normal hydrogen atoms, but it is much easier to initiate fusion reactions involving deuterium than hydrogen[2]. Brown dwarfs are thus able to generate energy, and emit light using deuterium rather than hydrogen fusion. Typical surface temperatures for brown dwarfs range from 2000 to 700 K, so their emission peaks in the near-infrared.

Brown dwarfs were first theoretically discussed in the 1960s, but it was many years before the first examples were discovered. This is because the necessary sensitivity to detect them in the infrared only became available much later. The first suspected brown dwarf was discovered in 1988 when an infrared survey looking for companions to white dwarf stars (a late stage, post-main-sequence end point for the evolution of a star with a mass roughly the same as that of the Sun) found a faint, cool companion to the star GD165 [13], which was dubbed GD165B. The surface temperature of this star, at 2100K, and derived mass, about 0.07 times the mass of the Sun, were both so low that they called for an extension of the usual stellar classification system. This object remained a curiosity until 1995 when two incontrovertible brown dwarfs, named Teide 1 and Gleise 229B were discovered, and confirmed by details of their optical spectra [139, 117]. The resulting theoretical models, and later discoveries of other brown dwarfs, extended stellar types beyond M

---

[2]This is one reason why experimental and, hopefully, future power generating fusion reactors on Earth use deuterium rather than hydrogen in their fuel supply.

to include three new classes of star, types L, T and Y[3] reaching temperatures as low as a few hundred degrees.

The key to finding the first brown dwarfs, GD165B, Teide 1 and Gleise 229B, was the first generation of sensitive near-infrared cameras. As infrared detector technology has improved (see Chapter 2) our capabilities at these wavelengths have expanded, and the number of brown dwarfs, and the range of brown dwarf properties, has expanded as well. While GD165B and Gleise 229B were found by pointing the telescope at specific targets - their companion stars - a proper characterisation of the population of brown dwarfs can only come from surveying large areas of the sky. These could be at random positions or in regions of active star formation. Since brown dwarfs are cooler than normal stars, an examination of the infrared colours of objects seen in such a survey can pick out candidate brown dwarfs which can then be followed up with more detailed observations by way of confirmation. Colour, in this context, means the relative amount of energy coming from a source at different wavelengths. A normal main-sequence star will be hotter than a deuterium-burning brown dwarf, so more of its energy output will appear at shorter frequencies. It will have, in astronomical parlance, a bluer spectrum - its emission peaks at shorter wavelengths - compared to a cooler brown dwarf, which will have a redder spectrum - its emission will peak at longer wavelengths. A demonstration of this can be found in Figure 4.9, and this kind of colour selection approach can be applied to finding a wide range of sources, as we will see later.

The first large samples of brown dwarfs emerged from the Two Micron All Sky Survey (2MASS) [154], a relatively shallow, i.e. not very sensitive, survey of the entire sky in the J, H and K infrared bands (wavelengths of 1.2, 1.6 and 2.2 microns). Selection of red, point-like objects from this led to the identification of several hundred L and T dwarfs [90]. The number of brown dwarfs found, bearing in mind that 2MASS can only detect them when they are relatively nearby, suggests that they are twice as numerous as stars of other spectral types. However, since they are very low in mass, they only contribute about 15% of the mass of other stellar types.

Deeper, more sensitive, infrared surveys from the ground have extended our catalogues of brown dwarfs still further. The UKIDSS (UK Infrared Deep Sky Survey) Large Area Survey [173], is somewhat smaller in area than 2MASS but can detect sources 25 times fainter. It has found brown dwarf stars as cool as 750 K [172]. More sensitive large area infrared surveys, using instruments such as the VISTA telescope, are already underway [7].

In space, the WISE satellite (Wide field Infrared Survey Explorer) [177] has conducted a survey of the entire sky at the somewhat longer wavelengths

---

[3]Why were these letters chosen? The answer seems to be that they are three of the few letters still available for classifying stars. As well as the main sequence types, O, B, A, F, G, K and M, other letters are used for some special types of star, such as carbon stars (see below), while others might cause confusion with other classes of object e.g. X for X-ray sources. L was chosen as the closest available letter to M, while T and Y were the last letters left in the barrel.

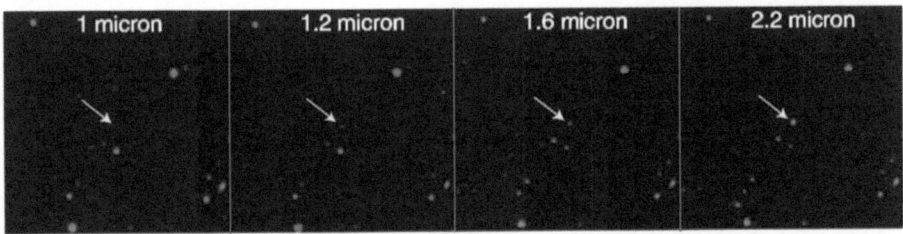

| 1 micron | 1.2 micron | 1.6 micron | 2.2 micron |

Figure 4.9 How to find a brown dwarf.
A series of images of a brown dwarf star, indicated by the arrow, at increasingly long near-infrared wavelengths, ranging from 1 micron (aslo known as the Y band), to 1.2, 1.6 and 2.2 microns (the J, H and K bands). The brown dwarf's spectrum peaks at longer wavelengths, so it can be more easily seen at longer wavelengths, in comparison to the other objects in this image which are more normal, main sequence stars. This is described as having a redder spectrum. (Courtesy of S. Warren, N. Skrzypek, unpublished data.)

of 3.4, 4.6, 12 and 22 microns. The shorter-wavelength channels allow brown dwarf searches to be extended to even colder, and thus redder, sources, than is possible from the ground. The WISE survey was only completed recently, with catalogs released in 2012. These include over 563 million sources, so we are still in the early days of exploiting this huge new database of astronomical objects. Nevertheless, some exciting new results on brown dwarfs have already emerged, including the detection of the coldest brown dwarf so far, a Y-dwarf with a surface temperature in the region of 250 to 400K - not far off room temperature [14] - and the discovery of a binary brown dwarf pair just 6 light years from the Sun [102]. These two brown dwarfs thus make up the third closest stellar system to the Sun, after Alpha Centauri (4.4 light years) and Barnard's Star (6 light years).

These brown dwarfs don't look like normal stars. They are sufficiently cold that clouds can form in the upper layers of their atmosphere, possibly making them appear more like gas giant planets, like Jupiter, than our usual picture of a star as a bright sphere with occasional small star spots. The Hubble Space Telescope has detected clouds in the atmospheres of two brown dwarfs through near-infrared spectroscopic monitoring [6]. This shows that the amount of light received from them changes with time in just the way you would expect if a small number of large cloud complexes were present in their upper atmospheres. The clouds seem to be associated with potassium, sodium and methane features, but their exact constituents are so far unclear.

4.6 WHEN IS A STAR NOT A STAR? WHEN IT'S A FREE FLOATING PLANET

Detailed studies of individual brown dwarfs provide insights into their physics and chemistry, allowing us to test models for the atmospheres of massive bodies in a number of different and previously inaccessible ways. They also begin to blur the boundaries between stars and planets. Indeed, it is unclear where the deuterium burning limit ends. Some have suggested [37] that there may be a very small amount of deuterium burning underway in Jupiter, and that this might explain the fact that Jupiter emits more energy than it receives from the Sun. Results from the study of brown dwarf atmospheres can also improve our understanding of the properties of exoplanets, and especially the hot Jupiters that have been found in surprisingly large numbers.

In probing the bottom of the main sequence, and thus the extreme low mass end of the IMF, we have to ask how low can the IMF go? Is there a limit to how small an object can form through the normal processes of star formation? If there is no limit, then we might expect to find large numbers of 'planets' the size of Jupiter, or even smaller, floating freely in interstellar space without the benefit of a parent star. These objects would have formed by processes similar to those that form stars, and unlike those that form planets in a pre-existing solar system.

As we have already discussed, rocky terrestrial planets in our own Solar System, like the Earth, lost their initial hydrogen and helium atmospheres because of heating by the Sun. This wouldn't happen to a free-floating planet, so we might expect them to look more like gas giant planets. Hints of their existence have already been found through their gravitational effects on light from background stars. This gravitational lensing effect (for more details see Chapter 8) leads to an increase in brightness when another mass passes close to our line of sight towards a background star. The time signature of these lensing events can be used to determine the mass of the lensing object even if we cannot directly see any light from it.

A search for these lensing events [159] found a surprisingly large number of them, consistent with free floating bodies of mass between three and fifteen times that of Jupiter, below the masses we expect for brown dwarfs. These results suggest that there may be as many free-floating planets in the galaxy as there are stars of all spectral types - maybe 200 billion in all.

Whether these free-floating planets formed in Solar Systems and were later ejected, or whether they formed on their own is unclear. While some planets are certainly ejected from their parent systems, observations in millimetre waves and the near-infrared [62] suggest the presence of small, compact gas clouds that will form free-floating planets in the star formation region known as the Rosette Nebula. (see Figure 4.5). A young free-floating planet, formed from such a 'globulette', has recently been found through combining optical and near-infrared observations [100]. This object, named PSO J318-22, is about 12 million years old and lies about 80 light years away from the Sun with a mass about 6.5 times that of Jupiter. While formally classified as an L-star, the properties of PSO J318-22 turn out to be quite similar to those of recently found young gas giant planets inside planetary systems. The line between high mass planets and low mass stars seems to be blurring, and the range of objects that form like stars and travel freely through space is getting ever richer.

4.7 THE REMNANTS OF PLANET FORMATION

The formation of planetary systems is, as we saw in the case of our own Solar System in Chapter 3, a rather messy business. As well as the planets themselves being formed, along with their moons and rings, a lot of stuff gets left over. In our own Solar System the debris left behind after planet formation and its early evolution is what makes up the zodiacal dust, the asteroid belt, the Kuiper belt and the Oort cloud. The first three of these form disks of material around the Sun, while the Oort Cloud, which lies much further away from the Sun, lies in a roughly spherical distribution.

The direct detection of extrasolar planets is very difficult, since they are outshone many times over by their parent star. Most exoplanets currently known were discovered by observing their effect on their parent star, whether this is the small reduction in the brightness of light we receive from the star when they pass in front of it, or by the reflex motion of their star as they orbit around a common centre of mass. Only a handful of extrasolar planets have been directly observed, but this will hopefully change with the advent of new instruments designed to eliminate the light of their parent star [107].

The direct detection of debris disks around other stars is a different matter entirely. While the mass of material in such a debris disk is small compared to the mass of even a terrestrial planet - maybe as much as 10% the mass of the Earth - this material is spread over a large area, rather than being compressed into a single planet, making it much easier to see. The emission seen from these debris disks around other stars is thermal radiation - the dust and small bodies that make up the disks absorb light from the parent star, and this heats the material a little, to temperatures of about 50 K. The debris disk particles then radiate this heat away as simple black bodies, producing an emission spectrum that peaks in the far-infrared. This means that observations at these wavelengths can detect debris disks like the Kuiper Belt around other stars. Material closer in to the parent star, like the asteroid belt or the zodiacal dust, is less easy to detect since, judging from our own Solar System, there is rather less material in these regions (the asteroid belt amounts to about 4% of the mass of the Earth's moon, while the mass of the zodiacal dust is that of a single asteroid) meaning that it will be much fainter. The Oort cloud, in contrast, might amount to several times the mass of the Earth, but this material is much further away from the Sun and so will be much cooler, making an exo-Oort cloud much harder to detect. Indeed, the evidence for the Oort cloud in our own Solar System is largely indirect, based on the properties of long-period comets.

Searches for possible extrasolar debris disks became possible with the advent of ground-based submilimetre cameras, such as SCUBA on the JCMT, and met with substantial early success with the discovery of a ring of dust and gas around the nearby star Epsilon Eridani [67]. The properties of this ring of dust are similar to what would be expected for a debris disk in a young solar system. Since Epsilon Eridani is less than a billion years old, this is entirely reasonable. Perhaps more interesting is the fact that a ring of material was detected, rather than a disk. The inner regions of the system are free of dust, suggesting that the material has been cleared away or collected by a forming planet. Further studies of this disk, seeing how structure within it changes with time, has confirmed the idea of a planet, and suggest that a rain of comets is falling into the inner parts of this system, producing conditions similar to those of the late heavy bombardment of Earth during the first billion years of our own Solar System.

Since this initial discovery of a debris disk, many others have been found around a variety of stars. Examination of their properties allows us to gain insights into the processes by which planets may, or may not, have formed in these systems. Large surveys of stars for debris disks are underway both from the ground and in space. The Spitzer and Herschel spacecraft, in particular, allow much more sensitive searches for debris disks, and allow the temperature and other dust properties of these systems to be determined more accurately. The DUNES survey [52] using Herschel has found debris disks around 24% of the 133 nearby, mature, Sun-like stars it observed, suggesting such disks are a common feature of other solar systems.

As well as allowing for large surveys, the detailed studies of individual systems made possible by modern instruments have also been very revealing. A good example of this is the debris disk around the star Fomalhaut, an A-class star, about twice the mass of the Sun lying about 25 light years away. This disk forms a ring around the star with a radius of roughly 133 astronomical units (1 AU is the distance between the Earth and the Sun, roughly 150 million km). The disk has been seen in the optical through scattered light from the star in observations from the Hubble Space Telescope, and in thermal emission by both Herschel and ALMA (Figure 4.6). The fact that the debris disk has a ring-like structure implies the presence of a planet that has cleared out the inner regions of the disk. However, the ALMA observations show that the ring is actually very thin, about 17 AU wide, and that the edges of this ring are very sharp. This suggests that there are in fact two planets responsible for the shape of the ring, one inside it, and one outside [19]. This is a configuration that is similar to some of Saturn's rings which are kept in position by shepherd moons orbiting just inside and just outside of the ring material. In the case of Fomalhaut we may be seeing this process in operation on a much larger scale, with gas giant planets on either side of the debris ring keeping it in position. The dust around Fomalhaut is something of a scientific goldmine, as there are now hints that an analogue of the Solar System's zodiacal dust - hot and warm dust close to the star, detectable in the near- and mid-infrared - is also present, closer to the star [112].

4.8 AGEING STARS, LOSING WEIGHT

Stars cannot keep burning hydrogen forever, since their fuel supply will eventually run out. Low mass stars, however, burn hydrogen so slowly that they can remain on the main sequence for a very long time. A 0.1 solar mass star, for example, can stay on the main sequence for as much as ten trillion years, many times the current 13.8 billion year age of the universe. More massive stars, including our own Sun, do not have such a long lifespan to look forward to.

As hydrogen burns to helium inside a star, helium, the ash of this reaction, gradually collects in the star's core. Eventually, there is so much helium there, that hydrogen fusion reactions cannot continue and the core begins to collapse under its own gravity. The hydrogen-rich regions of the star around the core collapse as well, getting hotter and denser as they collapse, until hydrogen fusion restarts in a shell around an inert, helium-dominated core. The increased temperatures and pressures in this hydrogen-burning shell lead to faster burning of hydrogen and thus an increased rate of energy generation. This heats the outer layers of the star more than before, puffing them outwards. The star expands to a much larger size, and its outer layers cool.

This produces a type of star known as a red giant. Figure 4.1 shows where they lie on the HR diagram - they are cool, but have a high luminosity because of their very great size. Energy from the helium-burning shell is carried to the bulk of the star by convection, so heavier elements, cooked by fusion in the inner regions of the star but never before released, are now distributed to the outer layers. This process will happen to the Sun in about five billion years time, with the Sun's expanding outer, red giant envelope likely absorbing the Earth and everything on it.

The material dredged up from the inner reaches of red giant stars will be rich in elements heavier than helium, including oxygen and carbon. The chemical signatures of these elements can be seen in the spectra of red giant stars. A good example of this is the star IRC+10216, the nearest red giant star to the Sun, at about 400 light years away. IRC+10216 is also classified as a carbon star since it has a large amount of carbon-rich dust in its outer envelope. Herschel observations have also shown that the outer envelope of this star contains water [45] at temperatures up to 1000K. The only way that water could be produced in this star is if ultraviolet light from space has been able to break up some of the carbon monoxide (CO) molecules that would otherwise have collected all of the oxygen in this object. As well as having complex chemistry going on in its outer envelopes, IRC+10216, like other red giant stars, is also unstable, and losing mass to interstellar space. This star pulsates, with a 649 day cycle, and has thrown off several shells of material

Figure 4.10 An old star shedding dust shells.
Image of the red giant carbon star IRC+10216 as seen by Herschel. The star
is surrounded by several shells of dusty material that have been thrown away
from its outer envelope. One of these can clearly be seen here to the left, and
there are others closer in to the star. (Courtesy of ESA.)

from its outer envelope - one of these can be seen in images taken by Herschel
in the far-infrared, demonstrating the presence of dust in these shells (Figure
4.10).

As hydrogen continues to burn in the shell, more and more helium collects
in the core. The core gets denser and hotter until, for stars half the mass of
the Sun or larger, the core reaches a temperature of 10^8 K (1 hundred million
Kelvin). At this point helium fusion can begin. For stars less massive than
about two and half times the mass of the Sun, the entire core ignites at once
in a what is called a helium flash. This can disrupt the star and lead to the
ejection of matter from its outer envelope. In more massive stars, helium fusion
ignites more gradually. The same process that we saw take place for hydrogen
burning then happens: the heavier element products of helium fusion, carbon
and oxygen, collect in an inert core, with helium fusion taking place in a shell

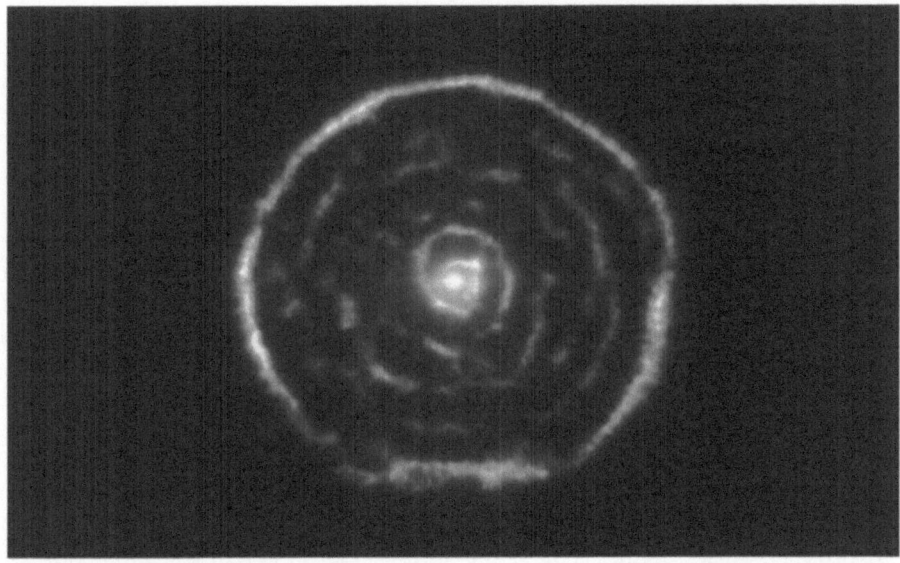

Figure 4.11 Dust shells around the AGB star R Sculptoris.
Observations of the dust shells around the AGB star R Sculptoris by the
ALMA telescope. (Courtesy of ALMA http://www.almaobservatory.org/.)

around it. Meanwhile, hydrogen fusion continues in a shell around the helium
fusing region.

Once a helium-fusing shell has become established, the star enters a second
giant phase, becoming what is called an asymptotic giant branch (AGB) star.
As with the red giant phase, these stars are variable over long periods, and
the stars lose a substantial amount of mass from their upper atmospheres.
In a similar manner to red giant stars, the convection that carries heat from
the helium and hydrogen burning shells to the outer layers also brings with it
heavier elements generated by fusion. This process is called the second dredge
up. Dust produced from these heavier elements is also lost from these stars.
Over time a series of dust shells spreads out from the AGB star into the
interstellar medium - an example of this can be seen for the AGB star R
Sculptoris in Figure 4.11.

As the AGB star ages, the helium-burning shell will run out of fuel. But
hydrogen burning, producing more helium, continues in a thin shell around
the, for the moment, exhausted helium shell. This generates more helium,
which, over timescales of 10,000 to 100,000 years, gradually accumulates until
there is enough to start fusion again. In a manner similar to the helium flash
at the start of the AGB phase, this produces a helium shell flash. The energy
produced by this flash is up to a thousand times the normal energy output of
the star, so the star rapidly expands, and cools. This drops the temperature
and density in the hydrogen burning shell to levels that cannot sustain fusion.

The star contracts, hydrogen fusion restarts and the process repeats, with another helium shell flash sending further thermal pulses out into the star. This stage in the life of an AGB is termed the Thermal Pulse AGB stage. Convection processes during the thermal pulses bring further enriched material into the outer layers of the star in what is termed the third dredge up. This leads to complex chemistry in the stellar atmosphere and in the dust and gas expelled from the AGB star during thermal pulses.

AGB stars in both the early and thermal pulse phases expel large quantities of dust into interstellar space, and are thought to be one of the main sources for the dust we see in galaxies. The term 'interstellar dust' covers material that has grain sizes up to a few microns - roughly the size of the particles that make up smoke - and down to what are essentially large molecules of polycyclic aromatic hydrocarbons (PAHs). The structure of a grain of interstellar dust can be quite complex, starting with a core of either silicate or carbon, around which accumulate atoms of elements common in the interstellar medium and in the outflows from the star that has produced the dust, including hydrogen, oxygen, carbon and nitrogen. These interact chemically to produce an icy mantle containing water, methane, carbon monoxide and ammonia. The surface of this dust mantle will also be irradiated by ultraviolet light in interstellar space, driving further chemistry, and producing a sticky outer layer.

Dust produced by these stars is a key component of the interstellar medium of all galaxies. It helps to cool the giant molecular clouds from which stars form, absorbs light from stars, reprocessing it into the far-infrared, and plays a key role in the formation of planets. Understanding the origins of dust in AGB stars is thus an important step in understanding a driver of many astronomical processes, and also in understanding what we see when we observe astronomical objects in the infrared.

4.9 STELLAR DEATH

The pace of stellar evolution speeds up as heavier and heavier elements are burned. A star like the Sun will spend about 10 billion years on the main sequence, but only a billion years as a red giant. The AGB phase will last less than a hundred million years. For the Sun, the helium-burning phase will produce a dense core of carbon and oxygen which will never reach the temperatures necessary to produce further fusion. The thermal pulses at the end of the Sun's AGB phase will eventually blow away most of the outer layers of the star, revealing a hot, degenerate carbon and oxygen core. Ultraviolet light from this core will ionise the ejected gas and help to accelerate it away.

The end result will be what is called a planetary nebula (PN). This name was originally given to them by William Herschel who thought they were part of the formation processes of planets. He was wrong, but astronomers are now stuck with the name.

The PN phase is brief but spectacular, as you can see from Figure 4.7, which shows the Cat's Eye Nebula, also known as NGC6543, as seen in the optical. Around the central star you can see a series of shells, which are in fact spherical bubbles of ejected gas and dust, projected onto the sky. The shells were ejected at intervals of about 1500 years in a fairly steady, repeatable process. The inner regions of the nebula, characterised by asymmetric shells and jets, are very different. These parts of the Cat's Eye nebula were ejected about 1000 years ago, when the behaviour of this star seems to have dramatically changed.

The life of a PN ends after about 10,000 years, when the remaining hydrogen around the now-naked stellar core is exhausted and there is no more ultraviolet light available to keep the nebular gas ionised. The gases in the nebula cool, expand away from the dead stellar core and fade away to invisibility.

What is left is the cooling remnant, a small, very dense body known as a white dwarf. These are made of material a million times denser than water, so, while they have masses comparable to that of the Sun they are typically only the size of the Earth. White dwarfs have no internal means of generating energy, so, unless they are able to accrete material from a binary companion star, they will gradually cool. Most white dwarfs are seen at typical temperatures of 8000 to 40,000 K. The coolest seen so far, found using infrared observations and a colour selection method similar to that used to find brown dwarfs, have a temperature of about 4000 K. The coldest of these stellar remnants would require about 8 billion years to cool to this temperature. Adding a few billion years to that age to account for the lifetime of the parent star implies a maximum age for the region of our Galaxy in which they are found of about 10 billion years [97].

The search for cooler white dwarfs can help determine the history of star formation in our own galaxy. The universe is about 13.7 billion years old, so white dwarfs from the earliest generations of stars will have temperatures of a few thousand degrees. Searches for them, using infrared surveys, are underway.

The eventual destiny of a white dwarf is to cool until, after trillions of years, it is at the same temperature as the cosmic microwave background. Such hypothetical objects have been called black dwarfs, but, since the universe is not yet trillions of years old, none have ever been found. This is the ultimate fate of a star like our own Sun. More massive stars, more than about eight times the mass of the Sun, have a different and rather more spectacular fate.

4.10 SUPERNOVAE

Stars which have a mass greater than about eight solar masses can maintain fusion in their cores to much heavier elements than carbon and oxygen. In the same way that a lower-mass star has a hydrogen-burning shell beneath which helium collects, and a helium burning shell beneath that, below which lies an inert carbon and oxygen core, massive stars have a complete set of nested layers starting with a hydrogen burning shell and ending with an inert iron core. Iron is the end point of a whole series of fusion reactions including hydrogen, helium, carbon, neon, oxygen and silicon. This is because it is the most tightly-bound atomic nucleus, so fusing iron nuclei to make heavier elements absorbs, rather than releases, energy.

As time passes the iron core gets larger and the pressures inside it increase. This cannot continue indefinitely. Eventually the pressure becomes too great, and one of several processes take place which cause the core to precipitously collapse, triggering a supernova explosion. The simplest process by which a supernova is triggered is the capture of electrons by protons, which produces neutrons and releases energy. At this point in stellar evolution the core is being supported against collapse by the electrons it contains[4]. Each electron capture removes an electron from the core, reducing its ability to support itself, which means there will be further electron capture reactions. This triggers a runaway collapse. The end result of this process is the conversion of the core into a neutron star - what is essentially a giant atomic nucleus about 20 km across, but containing about a solar mass of matter. Higher mass stars can have somewhat different routes to a supernova, and may leave a black hole remnant rather than a neutron star.

All of these collapse processes release a vast amount of energy, heating the parts of the star outside the collapsing core, and blowing it apart in an explosion that can be seen across intergalactic distances. The material in the fusion-burning layers around the inert core is also ejected, releasing enriched material into the interstellar medium. The energy release in the core also drives energy-absorbing fusion reactions, producing elements heavier than iron and spreading these, too, into the interstellar space.

A supernova explosion is certainly spectacular, but lasts only a very brief time. Supernova remnants, however, last a lot longer - many thousands of years. The Crab Nebula, for example, is the result of a supernova explosion recorded by Chinese astronomers in the year 1054. The progenitor star is thought to have had a mass about twelve to fifteen times that of the Sun. Figure 4.7 shows a far-infrared image of the Crab Nebula as observed by Herschel.

This image helped to solve a long-running problem in astrophysics: Where does the dust in the interstellar medium come from? We already know that mass loss from stars, and especially AGB stars, is responsible for some of the dust, but there was a dispute as to whether supernovae were producers or destroyers of dust. They might be producers since they release a large amount of material, rich in the carbon, silicon and other heavy elements that are the constituents of interstellar dust grains. They might be destroyers of dust since they heat the interstellar medium to high temperatures, potentially boiling dust grains to destruction. Figure 4.7 shows that the Crab Nebula contains filaments of dust in the densest parts of the ejected material. Two types of

---

[4]The electrons are supporting the core through what is known as electron degeneracy pressure, which is a somewhat complex result of combining quantum mechanics and thermodynamics.

dust are found: about a quarter of a solar mass of silicate dust, and about a tenth of a solar mass of carbon dust. These observations have made it clear that supernovae are dust producers, not destroyers, and that they can release prodigious amounts of dust during their explosions.

4.11   CONCLUSIONS

When we think of stars, we think of the little twinkling lights we can see in the sky at optical wavelengths with our own eyes. In this chapter we have seen that there is far more to the life of stars than can be seen by the human eye or purely in the optical part of the spectrum. By extending our reach into the infrared we can examine low-mass stars, and search for brown dwarfs that are potentially just as numerous as the stars we can see in the optical. We can search for free-floating planets, which challenge our definition of what stars, and planets, are, and we can examine the late stages of stellar evolution, when the stuff that made the stars is given back to the interstellar medium that floats between the stars, and from which they originally formed.

For stars like our own Sun, these late stages of stellar evolution are relatively sedate, but still produce spectacular objects like planetary nebulae. For high-mass stars the process is much more explosive, as they die in the fireworks of a supernova. In both cases, though, they release dust into the interstellar medium, one of the key things that we observe in the infrared. It turns out that dust is also an essential ingredient in star formation. So the death of one generation of stars provides the seeds to forming the next generation. Star formation is the subject of the next chapter.

CHAPTER 5

Young Heat

5.1   HOLES IN THE SKY

The night is warm, even at an altitude of 9000 feet. That's what you get during summer on the edges of the Atacama desert in Chile. I'm at the La Silla observatory, part of the European Southern Observatory. It's on a mountain in the foothills of the Andes, sitting between the coastal fogs along the edge of the Pacific, and the mountain storms of the true Andes. It has clear skies, and good weather and, since the first telescope was built here in the 1960s, it has grown so many observatory domes that you might think the summit was infected with an obscure kind of giant mushroom.

I am between observing runs. I've been working in the optical for about a week and now someone else has a few nights on the same telescope before it switches to a near-infrared instrument and I get to take some more data.

It's my first trip to such a good astronomical site in the southern hemisphere, so I use some of the time to look at a very different night sky than the one I'm used to in the north. The great arc of the Milky Way is far bigger and brighter. It arches over the sky like a great river of light.

But that river has some holes in it.

Scattered across the sky, along the track of the Milky Way, are patches of darkness, places that the light of the countless stars of the galaxy we live in cannot penetrate. Some of these patches have names, like the Coal Sack Nebula. The name comes from what it looks like - a big black blob hanging in the sky - but it's also a good description of what is causing these holes in the Milky Way. They're filled with dust, which is absorbing the light of the stars behind them, making them blacked-out regions against the rest of the star-filled skies.

5.2   INTRODUCTION

At the end of Chapter 3 we saw how the geography of our own Solar System provides the first hints as to the processes by which stars and planetary systems might form. Laplace realised that the alignment of the orbits of the planets to the rotation axis of the Sun meant that the Sun and planets formed together from a collapsing cloud of gas and dust. This result was based on the Newtonian mechanics of the 18th century, dealing with the conservation of angular momentum. Further progress in the study of star formation required deeper understanding of the behaviour of gas clouds, and the relationships between temperature and pressure, to allow us to work out how interstellar gas would respond to its own gravitational attraction. This required the development of thermodynamics, a subject that was at the core of 19th century physics.

In the early years of the twentieth century, James Jeans looked at the behaviour of isolated gas clouds in interstellar space, examining the conditions under which they would be stable, and under what circumstances they would collapse under their own gravity. A gas cloud collapsing under its own gravity, for example, will get hotter, leading to an increase in pressure which acts against any further collapse.

However, Jeans realised that there is a critical point beyond which pressure alone cannot stop a cloud of gas from collapsing. The larger the mass of a cloud, the smaller it is in size, and the lower its initial temperature, the more susceptible it will be to gravitational collapse. A high-density cloud of gas at a low temperature will collapse, and lead to the formation of a star. Low density, hot regions of gas will not collapse to form stars.

Our first step in understanding the formation of stars is to look at the fuel from which they will form, the gas and dust in the interstellar medium (ISM), and the range of densities and temperatures found there.

5.3   THE INTERSTELLAR MEDIUM

Space is not empty. While it is a very good vacuum, it is not a perfect vacuum, empty of all matter. Instead, the space between the stars in a galaxy is filled by something called the Interstellar Medium (ISM). Since hydrogen is the most common element in the universe, the ISM is predominantly made up of hydrogen. This can come in several forms. Firstly, it can come in the form we are most familiar with on Earth, molecular hydrogen, where each hydrogen atom is combined with another hydrogen atom to form H_2. Secondly, it can come in atomic form, H, where the proton that makes up the nucleus of a hydrogen atom is orbited by an electron. Finally, the hydrogen can be ionised, so that the proton and electron are separated.

The relative abundance of these forms of hydrogen in the ISM at any position in our galaxy depends on the physical properties of the ISM, which vary from place to place. The ISM also contains helium and other trace materials made up of heavier elements, recycled into the ISM through the processes of stellar evolution, which we looked at in the previous chapter. These trace elements, including dust grains and molecules like carbon monoxide (CO), can

Figure 5.1 The view of the Galactic plane from the La Silla
Observatory.
The dark clouds that obscure the visible light from parts of the Galactic plane
are the denser parts of the ISM. This is a very similar view to that described
in the introduction of this chapter. (Courtesy of ESO, photo by S. Brunier;
http://www.eso.org/public/.)

play key roles in star formation and also allow us to examine the properties
of the ISM where it might otherwise be difficult.

The label 'ISM' applies to regions with a widely divergent set of properties,
from supernova remnants, where temperatures can be measured in the millions
of degrees (see Figure 4.7) to dense clouds that have yet to collapse and form
stars, which can be as cold as 10 K. The density of the ISM also varies over
a huge range. Regions associated with circumstellar shells (see Figure 4.10),
where stars are blowing off their outer laters, can have densities as high as 10^{15}
atoms of hydrogen per cubic metre (that's a million billion), while the most
diffuse parts of the ISM, the hot intercloud medium, have densities a thousand
billion times smaller, at just 1000 atoms per cubic metre. For comparison,
there are over 10^{25} atoms per cubic metre in the air you are breathing, so
even the densest parts of the ISM are still ten billion times more rarefied than
the Earth's atmopshere.

Where the ISM is densest, the dust within it absorbs the starlight passing
through. This gives rise to the dark clouds scattered along the plane of our

own Galaxy. These are especially visible from the southern hemispher
the centre of the Galaxy is visible from there (see Figure 5.1). Where
is least dense ultraviolet light ionises the hydrogen in it and heats it to h
temperatures. This also can occur where denser regions of the ISM are close to
a hot, bright star. Once ionised, hydrogen in the ISM is transparent, making
the least dense parts of the ISM rather difficult to study. Denser ionised parts
of the ISM associated with hot stars, though, produce distinct objects known
as HII regions, where HII indicates that hydrogen is in its ionised state
viewed in the optical, HII regions have a distinctive red colour (see Fig
This comes from a specific hydrogen emission line, known as Hα, that results
from some fraction of the ionised hydrogen in the HII region encoun
and recombining with, an electron to form an atom of hydrogen.

Since the densest parts of the ISM are opaque to the light of stars
have to investigate them at longer wavelengths which are less susceptible to
absorption. This can involve observations of neutral, atomic hydrogen at radio
wavelengths using an emission line at a wavelength of 21 cm. In the den
est parts of these clouds, obscured from the outside by dust associated with
the gas, molecules of hydrogen can form. Outside dense ISM clouds
molecules would be rapidly broken apart by ultraviolet light from stars
this light is all absorbed in the outer parts of the clouds. Other molec
such as carbon monoxide (CO) and hydrogen cyanide (HCN), can als
inside these clouds. CO is the second most abundant molecule in the un
verse, the most abundant being molecular hydrogen. CO lines can be
by relatively low temperatures, making them very useful as a tracer of cold
material. Emission lines from molecular hydrogen correspond to much
temperatures and so are less useful for studies of cold gas. Molecules like CO
and HCN can be seen deep inside even the most obscured clouds since
emit in the far-infrared and submillimetre, making them largely immune t
the effects of obscuration. Emission lines from these molecules are u
trace the properties and behaviour of the densest parts of the ISM. And it is
these densest regions that are of most interest in the study of star formation
because, as Jeans told us (see Section 5.2), it is the densest, coldest gas that
is most susceptible to gravitational collapse.

---

[1]Astronomers use a somewhat archaic way of labelling different spectra. Where a chemist
would refer to spectral lines coming from ionised hydrogen as being associated with an H
ion, astronomers refer to the 'second spectrum of hydrogen', i.e., HII, which is the same
thing. This, of course, confusingly means that the first hydrogen spectrum, HI, comes from
neutral, un-ionised, hydrogen. Similarly OII lines come from singly ionised oxygen, O
OIII lines come from doubly ionised oxygen, O^{++}. This can make discussing things with
chemists a bit confusing!

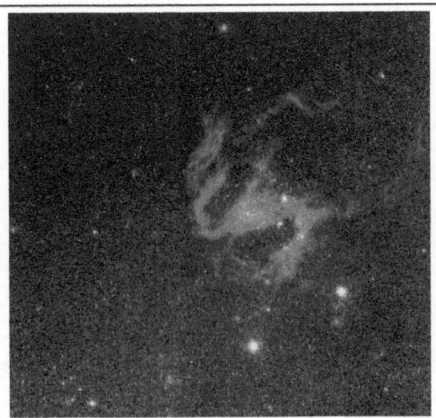 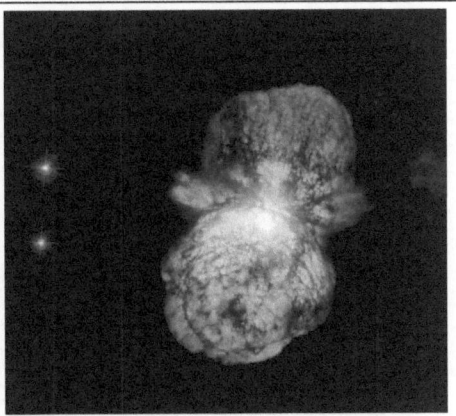

Figure 5.2 Left: The HII region N119 in the Large Magellanic Cloud. Right: An optical image of the object Eta Carinae.

Left: The diffuse emission seen in this image is emission from ionised hydrogen in the HII region N119. Courtesy of ESO. Right: Eta Carinae is a multiple star system that includes one of the most massive stars in our galaxy. What you see here is a nebula of gas and dust expelled by the massive star which may be on the way to exploding as a supernova. This is optical imaging from the HST. (Courtesy of NASA.)

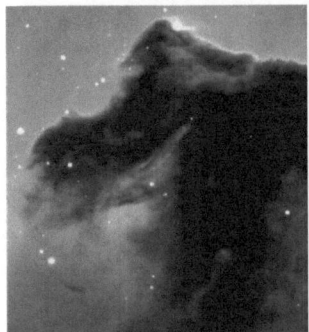 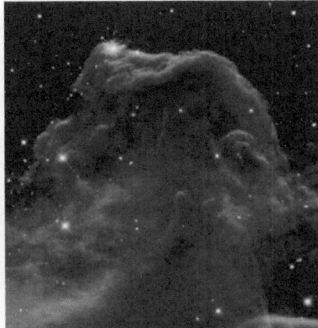 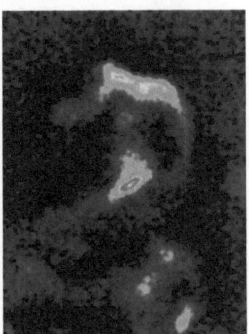

Figure 5.3 The Horeshead Nebula, as viewed in the optical (left), near-infrared (middle) and submillimetre (right).

Views of the Horsehead Nebula show different things at longer wavelengths, with the interior only visible at the longest wavelengths. These reveal what may be a pre-protostellar cloud, which may later collapse to form a star. (Left: Courtesy of ESO http://www.eso.org/public/; Middle: Courtesy of NASA; Right: Courtesy of D. Ward-Thompson.)

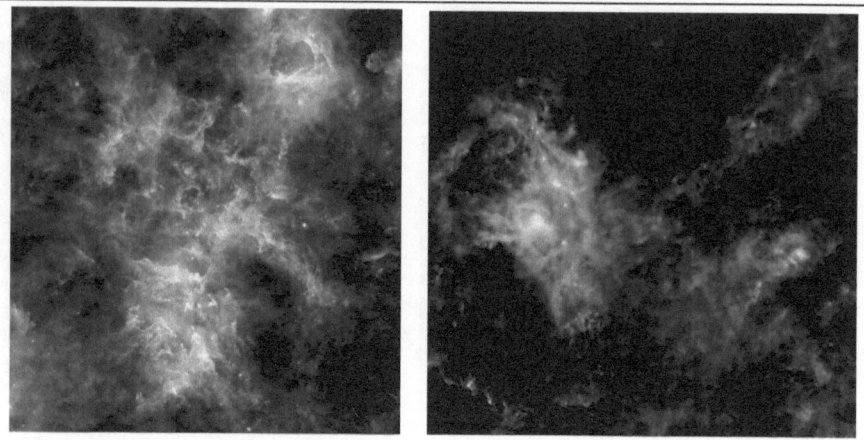

Figure 5.4 Left: Carina Nebula. Right: Aquila star-forming region.
Dust images from Herschel with blue showing 70 micron emission, green 160
micron and red 250 micron. Eta Carinae sits right at the centre of the image
to the left, heating the closest dust to higher temperatures, which appear blue
[136]. (Courtesy of ESA; http://www.esa.int/ESA.)

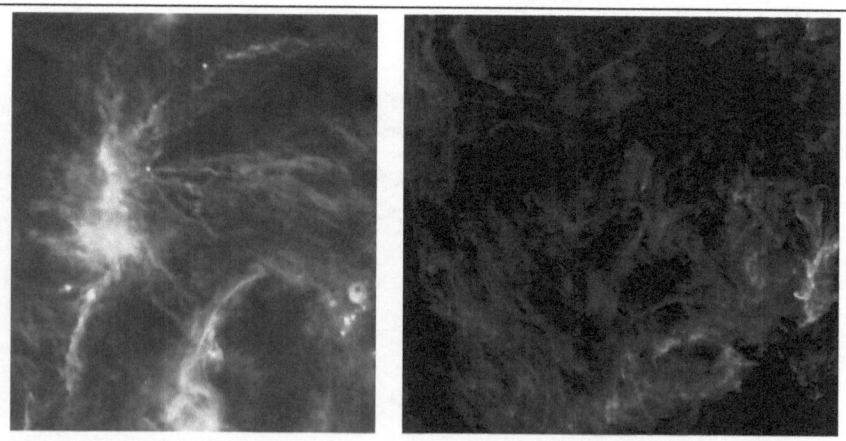

Figure 5.5 Filamentary structures found in the interstellar medium by
the Herschel Space Observatory.
Left: The DR21 region where massive star formation is underway showing the
filaments where stars are forming, and sub-filaments feeding more gas into the
densest regions. Right: An image of the Polaris region also shows filaments
in the interstellar medium, but there is no star formation underway here.
(Courtesy of ESA; http://www.esa.int/ESA.)

5.4 FROM GIANT MOLECULAR CLOUDS TO COLLAPSING CORES

The ISM can have a rather hierarchical structure with larger clouds of diffuse material having smaller, and denser, clouds inside them. The largest clouds in the ISM are known as giant molecular cloud complexes (GMCs). These can be up to 300 light years across and have masses up to a million times the mass of the Sun. Buried within these are the dense clouds and denser cores and filaments of gas that can become the sites of star formation. An example of a GMC can be found in Figure 4.4, which shows the Herschel image of the W3 GMC. In this image you can see the complex structures and substructures that exist within a GMC.

Jeans' discussion of collapsing clouds of gas was based on the idea that a spherical cloud of gas can collapse quietly and calmly on its own, without any disturbances from outside. Given the much more complex substructure of GMCs that we observe, it is clear that this is an oversimplification. But the more complex, real environments of GMCs offer plentiful opportunities for gravitational collapse of the kind Jeans envisaged to be triggered by outside events. Possibilities for these include the following:

● The impact of a blast wave, or shock front to use the more technical term, from a nearby supernova.

● The impact of shocks from other causes, such as the presence of nearby bright, hot, young stars. You can see hints of this in the Herschel image of W3 (Figure 4.4), where hotter dust (shown in blue) can be seen in the top left. HII regions are often produced by hot, bright, young stars, so star formation itself can be a trigger for more star formation. An example of this is seen in Figure 5.6, which shows part of the HII region known as IC2944. The dark patches are dense clouds of gas and dust known as Bok Globules, some of which are known to host forming stars. These kinds of shocks don't always assist star formation but can in some circumstances hinder it as well, making things rather more complicated.

● Density changes resulting from the structure of our own galaxy. The spiral structure of our own Galaxy (see Chapter 6) is a result of waves of increased or decreased density sweeping around it. Not surprisingly, regions of active star formation are associated with the denser regions of this spiral structure.

● Gravitational interactions between clouds inside a GMC. These can lead to density enhancements. Smaller clouds can also collide and merge, triggering gravitational collapse and the formation of stars.

● Gravitational interactions and mergers between entire galaxies. As we will see in Chapter 6, these can trigger star formation on a very large scale.

Figure 5.6 The IC2944 HII region and the Bok Globules within it.
IC2944 imaged in the optical by the Hubble Space Telescope. (Courtesy of
NASA.)

Once a dense cloud core starts to collapse, the Jeans process takes over and it will begin to fragment into smaller, collapsing cores that will eventually form stars. This explains the tendency for stars to form in groups or clusters.

5.5   PROTOSTARS TO STARS - THE BIG PICTURE

Once a dense cloud fragments into separate clumps of gas that are collapsing under their own gravity, the process of star formation has begun, and the clump is on the way to becoming a protostar. The first phase of star formation is the pre-protostellar core collapsing under its own gravity. These collapsing cores are cold, typically just 10-20K, and have low enough densities that they don't heat up as they contract. Once this stops being the case, and the densest region at the centre of the pre-protostellar core stars to get hotter as well as denser, then the object is a protostar, and star formation is underway.

The process through which a protostar develops into a normal, main sequence star, goes through four key stages, leading to the classification of protostars into four classes: classes 0 to 3 (see Figures 5.7 and 5.8). Originally, there were only three classes of protostar, class 1 to 3, until astronomers working in the far-infrared and submillimetre identified objects that represented a key, earlier stage in the process [2], which was labelled class 0, since class 1 had already been taken. The pre-protostellar collapse phase of star formation can last up to a million years.

A class 0 protostar will have a low mass core that grows by accreting material from a disk of circumstellar material that forms around it. This disk will be embedded in the remnants of the pre-protostellar cloud, which will continue to collapse onto the disk. There will be plenty of dust and gas in this collapsing cloud, which will still be very large - perhaps a few thousand astronomical units in size[2]. This means that emission from the young stellar object (YSO) at the core will be difficult to see at wavelengths shorter than the mid-infrared. There will also be outflows of material from above and below the disk. These result largely from the heat of the YSO boiling material off the surface of the accretion disk, but there may also be a contribution to these outflows from the growing magnetic field of the YSO. These bipolar outflows help to remove angular momentum from the accretion disk and accelerate the rate at which material from the disk is accreted onto the protostar. Unlike the pre-protostellar phase, which can last a million years, a forming star passes through the class 0 phase quite quickly, in less than about 30,000 years. The brevity of this phase, coupled with the fact that class 0 protostars can only be found at long wavelengths, in the mid- and far-infrared and radio, accounts for the difficulty astronomers studying star formation had in finding them.

The class 1 protostellar phase lasts rather longer, about 200,000 years. In this phase the YSO has grown to have a mass greater than that of the embedding accretion disk and envelope, unlike the earlier class 0 phase when

---

[2]An astronomical unit is the distance from the Earth to the Sun, about 150 million km.

the envelope was more massive. This means that the YSO is less obscured by surrounding dust and gas, and can now be seen in the near-infrared, though the majority of its energy output still comes in the mid- and far-infrared. The accretion disk remains well developed, though the system has now shrunk to a size of a few hundred astronomical units. The bipolar outflow remains, but, as a result of the dwindling effects of the shrinking envelope, this outflow is now less of a jet and more of a broader, less collimated, spray.

The end of the class 1 protostar phase marks the start of the protostar's life as what is generally thought to be a star. These objects are yet to start full hydrogen fusion, though, so they are not yet part of the stellar main sequence. Instead, their heat comes from the release of gravitational potential energy as they continue to contract and, at later stages, from burning other, less common, elements that are easier to fuse, such as lithium and deuterium. Class 2 and class 3 protostars (see Figure 5.8) are also known as T-Tauri stars, named after their prototype, the star T-Tauri. They are highly variable in their output, have strong emission lines in their spectra, and strong stellar winds.

Class 2 protostars, in the T-Tauri phase, still have disks of protostellar material around them, containing gas and dust amounting to about 1% of the mass of the central star. A young star spends about a million years as a type 2 protostar. During this time, the material in the remnant disk surrounding the star will itself develop its own accreting masses, which will gradually form planets (see the next section). The dust in this protoplanetary disk is heated by the central star to temperatures of a few hundred K, leading to significant emission in the mid- and far-infrared.

Once planet formation is complete, and the surrounding disk of remnant material has dropped to a mass equivalent to that of the planet Jupiter or less, then the protostar is class 3. This phase lasts about 10 million years. At the end of this period, the central star is finally dense and hot enough at its core to initiate full scale hydrogen fusion, and the star begins its proper life on the main sequence.

5.6 PLANET FORMATION

Planet formation occurs towards the end of the star formation process, within the protoplanetary disk of a class 2 protostar. If you compare these sorts of systems with the basic scheme that Laplace outlined in the 18th century, and that we discussed in Section 3.9, you'll see that it has all the necessary ingredients. There's a star at the centre, spinning in the same direction as the protoplanetary disk, and the chance for planets to condense out of this disk. In some ways the formation of planets from a protoplanetary disk is rather similar to the formation of a star - self gravitating bodies in a dense medium attract material onto them and grow over time.

However, the process by which planet formation starts is rather different. It is not gravitational, but driven instead by the tendency of dust particles

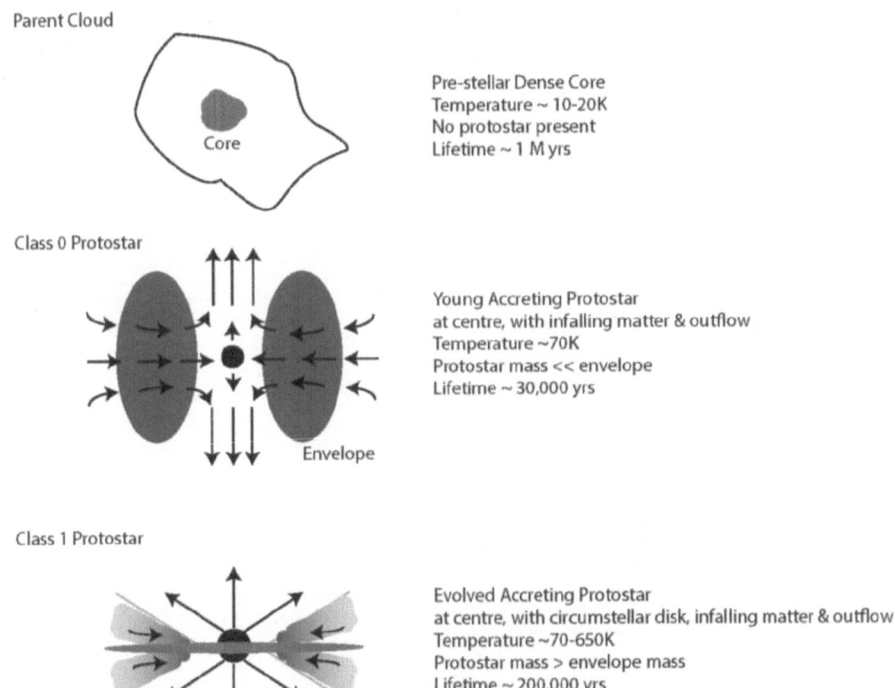

Parent Cloud

Core

Pre-stellar Dense Core
Temperature ~ 10-20K
No protostar present
Lifetime ~ 1 M yrs

Class 0 Protostar

Envelope

Young Accreting Protostar
at centre, with infalling matter & outflow
Temperature ~70K
Protostar mass << envelope
Lifetime ~ 30,000 yrs

Class 1 Protostar

Evolved Accreting Protostar
at centre, with circumstellar disk, infalling matter & outflow
Temperature ~70-650K
Protostar mass > envelope mass
Lifetime ~ 200,000 yrs

Figure 5.7 Diagrams showing the first three main stages of protostar evolution.
The evolution of a protostar from its parent cloud through class 0 to class 1 protostar. At the end of these stages, the young protostar is considered a pre-main sequence star. (Adapted from an original by P. André [3].)

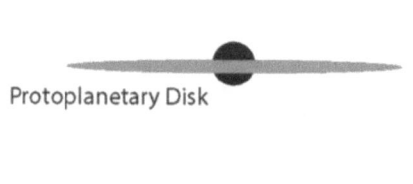

Protoplanetary Disk

T-Tauri Star
Much reduced accretion & outflow rates
T=650-2880K
Disk mass ~ 1% star mass
Lifetime ~ 1M yr

Class 3 Protostar

Debris disk and planets?

Weak T-Tauri Star
No outflow or accretion
T>2880K
Disk mass < Jupiter mass
Lifetime ~ 10 M yr

Figure 5.8 Diagrams showing the final two stages of protostar evolution. The evolution of a protostar through its pre-main sequence phases as a T-Tauri star. At the end of these stages hydrogen burning has started and the star has joined the main sequence. (Adapted from an original by P. André [3].)

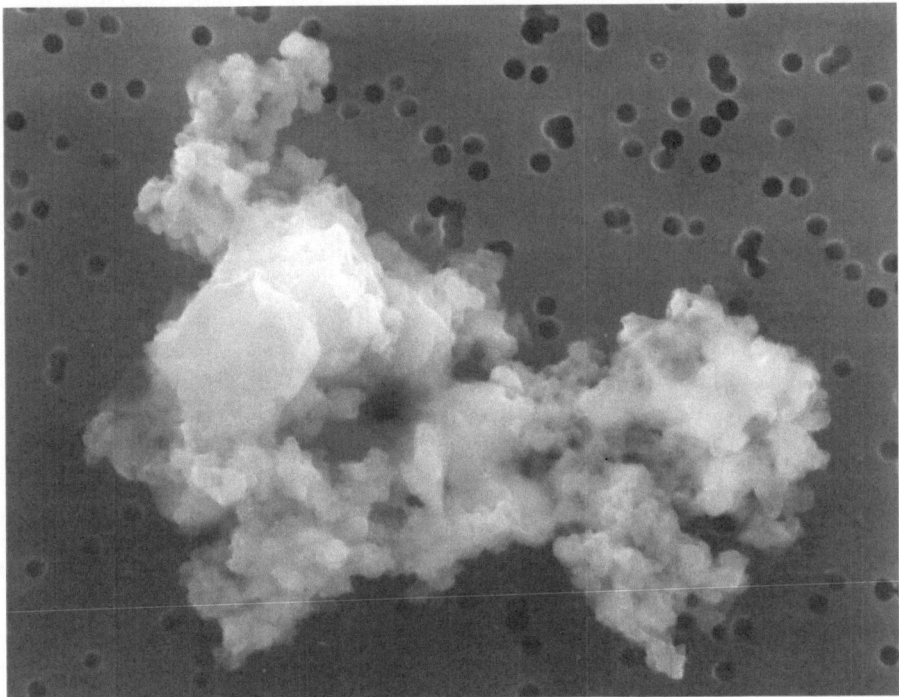

Figure 5.9 An interplanetary dust grain, collected by a NASA U2 aircraft.
(Courtesy of NASA.)

to stick to each other. An example of an interplanetary dust grain, collected from the stratosphere by a high flying NASA U2 plane, and probably quite similar to those that stuck together to start the formation of planets in our own Solar System, is shown in Figure 5.9. As you can see, this dust grain seems to be made up of lots of smaller dust grains which have stuck together. This adhesion effect is chemical or electrostatic in nature, since the grain is far too small for gravity to play a significant role in attracting its constituents.

Given sufficient time, clumps of dust grains like this, sticking to one another in the protoplanetary disk, will grow big enough for gravitational attraction to become significant. At that point, when the clumps are about 1km in size, they can be thought of as planetesimals, and have become the building blocks of planets.

5.7   PROTOSTARS: THE OBSERVATIONS

Our picture of how some dense interstellar dust clouds collapse to form stars, and the planets around them, was not drawn up in a purely theoretical manner. Instead it was inspired and has been thoroughly tested by astronomical observations. Since star formation involves dust, and many of its stages, especially the formative, early stages, take place in dust enshrouded environments, many of the observations that have led to our current picture of star formation had to be conducted in the infrared and submillimetre wavebands.

The starting point for star formation, and the home of pre-protostellar cores and class 0 protostars, are the densest parts of giant molecular clouds. These will appear as very dark clouds in the optical, some of which can be seen in Figure 5.6. Embedded inside these dark clouds will be pre-protostellar cores, in either the earliest stages of collapse or poised to collapse if some nearby triggering event, such as a local supernova, tips them over the edge of stability.

Figure 5.3 shows one of the more distinctive dust clouds in the local Galaxy, the Horsehead Nebula. In the optical this appears as the familiar dark equine silhouette, obscuring the light of background stars and the red nebulosity of the Orion HII region in which it lies. In the near-infrared, you can begin to see through the obscuring material to stars within and behind the head. You can also see emission from the clouds of dust and gas that make up the nebula, some of which comes from the gas itself, while some of what you see is reflected light from nearby stars. We can really only see what is inside the Horsehead Nebula by working at much longer, far-infrared and submillimetre wavelengths. This reveals strong dust emission in both the 'head' and 'neck' of the Horsehead, showing that dense, cold (temperatures of about 20 K and masses equivalent to a few times the mass of the Sun) clumps of gas and dust are lurking there. The properties of these clouds broadly match what we expect of pre-protostellar cores. The clumps of material inside the Horsehead are in fact probably not in the process of collapsing to form stars at the moment, but they may be triggered into star formation in the future under the influence of the nearby, hot young stars that are ionising the more diffuse hydrogen gas in the space around the Horsehead Nebula.

Class 0 protostars are deeply obscured by the dust and gas they are accreting from the surrounding pre-protostellar cloud, so the protostars themselves are best observed in the infrared or, in some circumstances, in the radio. However, a key feature of class 0 protostars are the jets of material boiling off their accretion disks. These form powerful outflows of material moving away from the poles of the forming star and can often be seen outside the obscuring cloud of material over a range of wavelengths, but usually in the infrared since they are still embedded in obscuring dust clouds (see Figure 5.10). Such outflows actually occur throughout the first phases of star formation. Sometimes they move so rapidly that changes in them can be seen from one year to the next[3].

As star formation progresses, the obscuring disk gets thinner, and the surrounding cloud gets smaller as more and more material is either accreted onto the star or expelled by the outflows. In a class 1 protostar, the forming star can be more readily detected in the near-infrared, while the outflow, less constrained by the remnants of the protostellar cloud, is less well defined. In Figure 5.11 you can see what one of these looks like as viewed by the Hubble Space Telescope in both the optical, using the WFPC2 camera, and the near-infrared, using the NICMOS camera. As you can see the near-infrared image is less affected by obscuration, with the obscuring disk seeming to be thinner and some possible hints that the star itself has been detected at the crook of the V-shaped emission. In contrast, the outflow material is much more prominent in the optical, with strong emission running vertically down the image. This is at least partly because the ejected material is emitting strongly in certain emission lines visible in the optical. A further example can be seen in Figure 5.12, which shows a somewhat older class 1 protostar that lies in the Orion star formation region.

The obscuring material is mostly eliminated by the time you reach a class 2 protostar. Instead, what is left is a narrow disk of material around the new, young, star, and the end of significant outflows. Two examples of this are shown in Figure 5.13. The young stars within these disks can be quite active, emitting strongly in the X-ray and radio as they work up to being capable of fusing hydrogen.

The end stage of star formation is an object with a remnant protoplanetary disk, or debris disk, with a young, late-stage T-Tauri star at its centre. Debris disks can, in principle, persist at some level long into the main sequence lifetime of a star. At some level the asteroid belt and Kuiper belt in our own Solar System are weak debris disks. Similar disks have been found around other stars, as seen in Figure 4.6.

5.8   GROUPS AND BINARIES

Stars are not always born alone. In fact most stars start out their lives in stellar groups or clusters, and many stars are in multiple systems, orbiting around one another in pairs, known as binaries, or in more complex systems. In fact the nearest stars to the Sun, Alpha-, Beta-, and Proxima-Centauri, all orbit around each other in a trinary system.

The presence of stars in groups is easily explained by their formation from the same progenitor molecular clouds. As these clouds collapse under gravity the details of the starting point of the collapse can determine whether the cloud fragments into one, two or more protostellar clouds. External influences, such

---

[3]A link to a movie showing such motion from a young star, somewhat more evolved than class 0, can be found on this book's website.

Figure 5.10 The class 0 Protostar L1527 as seen by Spitzer.
A class 0 protostar as seen in near-IR light by Spitzer. What you see here is light that has escaped from the protostar into a surrounding cavity in the host molecular cloud. This cavity was produced by outflows from the protostar. (Courtesy of NASA.)

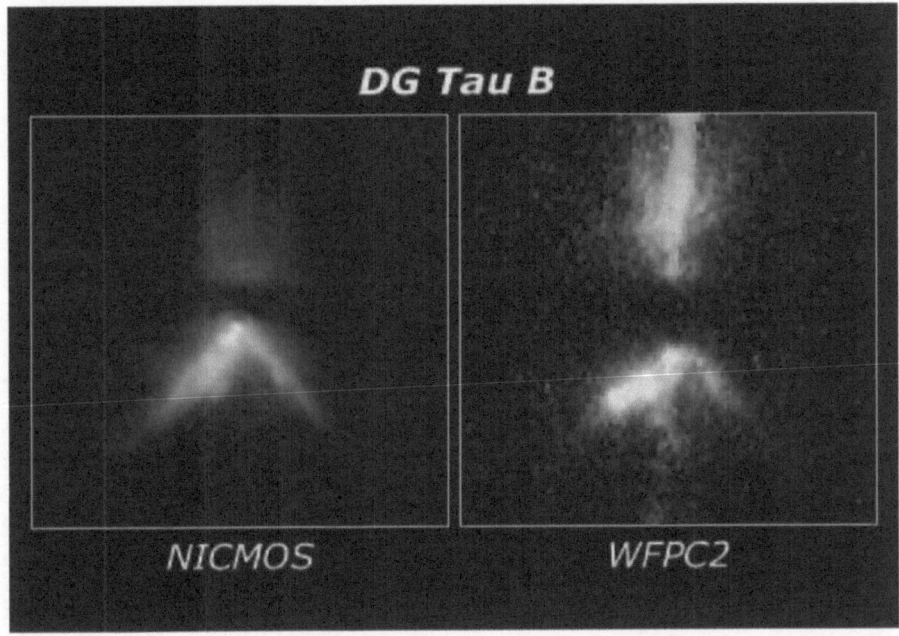

Figure 5.11   The class 1 protostar DG TauB imaged by the Hubble Space Telescope.
A class 1 protostar as seen in optical (right) and near-infrared light (left), observed by the Hubble Space Telescope. The star itself can just be seen in the infrared. (Courtesy of NASA.)

Figure 5.12   A forming star in the Orion star formation region given the designation 473-245.
The protoplanetary disk in this system is edge on and can be seen through its obscuration of light from the young star. (Courtesy of NASA.)

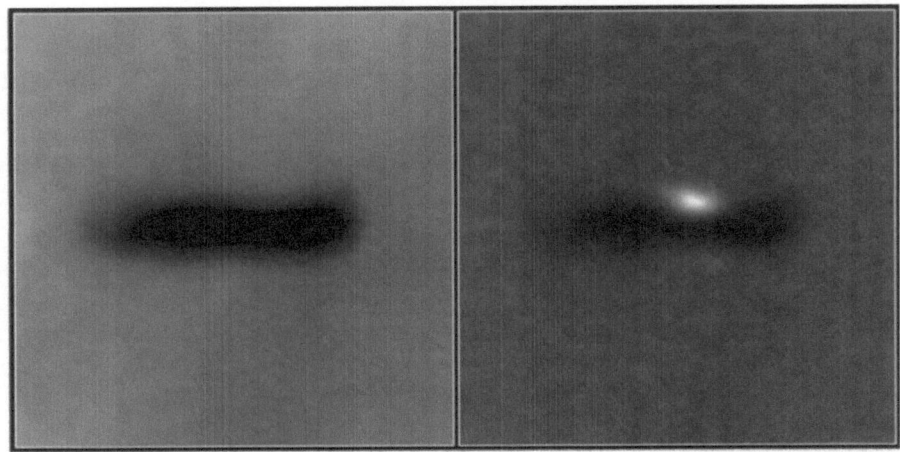

Figure 5.13 Protoplanetary disks viewed edge on in the Orion star formation region.
These are optical images from the Hubble Space Telescope. (Courtesy of NASA and ESA.)

as the presence of a nearby supernova or luminous star, can also affect the way a collapsing cloud fragments. The end results of this kind of fragmentation are usually clusters of stars. These are not gravitationally bound, and they eventually evaporate as member stars drift away over time. The Pleiades, see Figure 4.5, is an example of a fairly large group of stars known as an open cluster. It is one of the nearest and most familiar open star clusters to the Sun. Larger star clusters called globular clusters are also found. These are gravitationally bound, and can contain hundreds of thousands of stars which circle the Galaxy in a single cohesive structure. The globular clusters we see in our own Galaxy contain very little gas and dust, and no star formation. The stars in them are old, suggesting that, in the case of our own Galaxy at least, no new globular clusters have formed for many billions of years. Globular clusters and globular cluster-like structures can be detected in many other galaxies, some of which may be forming stars very rapidly. The circumstances which may drive the formation of this kind of object are discussed in Chapter 6.

The formation of binary stars and more complex systems may well result from something similar to the formation of stellar clusters, but on a much smaller scale. While an open cluster forms when a larger cloud fragments into separate collapsing protostars, a binary results when a collapsing cloud fragments on a smaller scale, leaving the separate components orbiting around each other. If this is the case, then the individual stars in binary pairs should have very similar ages, and this is in fact what is found (see [174] and further references in [157]). The alternative hypothesis, that binary stars form by gravitational capture, in much the same way that the planet Mars captured its moons Phobos and Deimos from the asteroid belt, would not lead to the ages of binary stars being the same.

5.9 HIGH MASS STARS

The foregoing discussion of star formation is based on observations and theoretical models of nearby star formation regions. It works well, and there is plentiful observational evidence to support the models, some examples of which have been shown. However, massive stars are much rarer than stars a few times the mass of the Sun, so there are no high mass stars forming sufficiently nearby that they can be studied in detail. The theory of star formation described above seems to work for stars up to about eight times the mass of the Sun, but beyond that there are problems.

The main difficulty in forming the highest-mass stars, which would be classified as O and B type stars, is that to reach such a high mass, the young protostar has to accrete enough of the necessary material. A young, high mass star will start fusing hydrogen into helium long before it has reached its final mass [157]. Once this happens, the young star will heat up and emit plentiful electromagnetic radiation. This radiation will exert pressure on the material falling towards the star, whether in the accretion disk or in the larger, more diffuse, surrounding protostellar cloud, counteracting the inwards pull of gravity. A large enough young star will exert enough radiation pressure on the infalling material that accretion is completely turned off. In fact, rather than falling onto the young high mass star, any remaining protostellar material will be blown away from it, halting any further growth.

On this basis you might expect there to be a maximum mass for stars, but that does not seem to be the case. Figure 5.2, for example, shows an image of the stellar system called Eta Carinae. This is now known to be a multiple star system [85], but the most massive member of this system is thought to have a hundred times the mass of the Sun [44], making it a true monster among stars.

How could such a star form if a much smaller star, just ten times the mass of the Sun, terminates its own growth through radiation pressure?

One hint is the fact that the most massive stars, O and B types, are found in groups or associations with each other. These associations are sufficiently common that they have a name, OB associations. And, since O and B type stars do not live very long, OB associations and other massive stars are usually found close to the star formation regions where they formed. Eta Carinae is a good example of this, sitting inside the Carina Nebula (Figure 5.4), a giant molecular cloud that is also home to HD93129A and HD93129B, two other very massive stars [119]. The Carina nebula is also notable for being very massive itself, containing gas and dust amounting to a mass roughly a million times greater than that of the Sun. This makes it four times as massive as the more nearby, and more famous, Orion nebula.

The trick with making massive stars is piling matter onto them faster than they can blow it away with the force of their own energy output. This is difficult to do if the material falling onto the young star is simply gas being accreted through the normal processes behind star formation. However, in a large, dense star-forming cloud like the Carina Nebula, you will find many stars in the process of formation. The closer these forming stars are to each other, the more likely that they will interact with each other gravitationally. In most cases such an interaction will result in the formation of binary stars, or more complex gravitationally-bound systems. In some cases, though, the stars will actually merge with each other. This bypasses the effects that would blow diffuse gas away from the merging stars, and allows the assembly of much more massive objects. If a succession of young stars or pre-stellar objects merge in this way, objects as large as Eta Carinae can be produced. Needless to say this stellar merger process can be complicated, but, at the moment, this seems the most likely way for massive stars to be assembled.

5.10 THE ORIGIN OF THE INITIAL MASS FUNCTION

The origin of high mass stars is just part of a broader problem in star formation, because, even though we now have a fairly good understanding of how pre-stellar clumps collapse and form individual new stars, we do not know what produces the overall mass distribution of stars, the stellar initial mass function (IMF), as described in Section 4.4.

The shape of the stellar IMF could come from the detailed physics of collapsing protostars, but there have long been suspicions that it has its origin much earlier in the star formation process, that it has something to do with the pre-protostellar cores and the molecular clouds from which they condense. These cores, as we have seen, are very cold and dense. They are opaque to optical and near-infrared light (see, e.g., Figure 5.3), and their temperatures are so low that they really only become prominent emitters in the far-infrared and submillimetre. The cores are also relatively small in size, about 0.3 light years across, so you need a telescope with good resolution to be able to distinguish the cores from the clouds in which they are embedded. A far-infrared telescope with those capabilities only became available with the launch of the Herschel Space Observatory - in fact one of its central scientific goals was to examine the nature of star formation and the origin of the stellar IMF.

Among the first images from the Herschel Space Observatory to be scientifically analysed was that of the Aquila star-forming region, shown in Figure 5.4 [4]. This data was used to detect and characterise the protostellar cores that lie in the dense filaments of gas and dust that fill the Aquila region [91]. It allowed the mass of each of the protostellar cores to be calculated and, since over 500 were found, it was possible to measure the distribution of these masses. This is plotted in Figure 5.14.

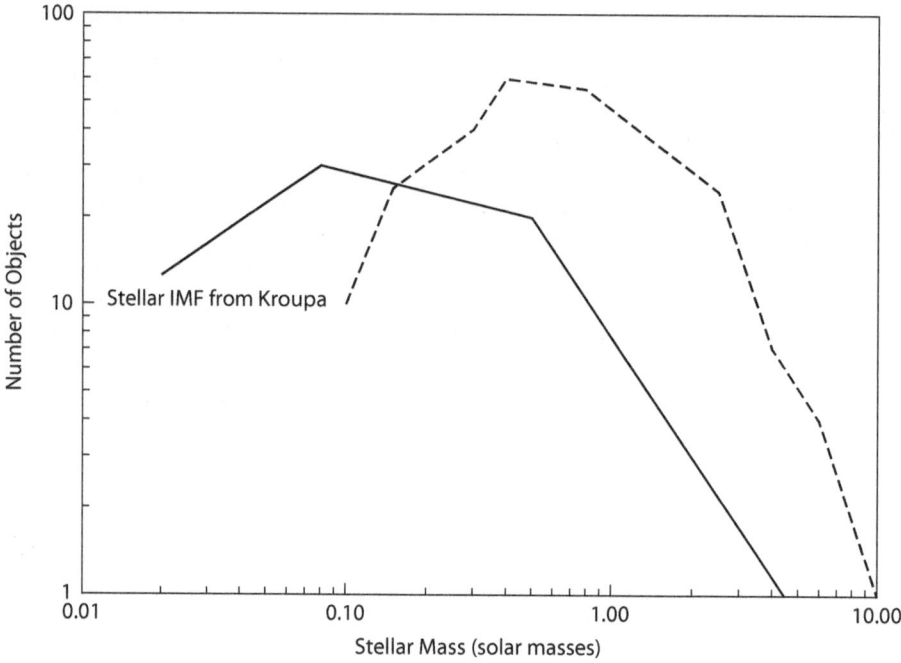

Figure 5.14 The clump mass function from Aquila compared to the stellar IMF.

The IMF as described by Kroupa [92] compared to the clump mass function as measured by Herschel in the Aquila region. The shape of the two functions is surprisingly similar.

As you can see, the shape of the stellar IMF and the prestellar clump mass function are very similar, but with the peak of the clump mass function lying at masses a factor of about three to five times greater than the peak mass in the stellar IMF. This suggests that the clumps being seen here are indeed the progenitors of stars, and that the mass distribution of clumps is what defines the mass distribution of stars. The offset between stellar and clump mass functions implies that star formation is an inefficient process, with less than half the mass of the initial clump ending up forming the star that collapses from it [1]. Given that we already know that an awful lot of matter is thrown out of protostars during the formation processes, this low efficiency is not a great surprise.

If the stellar IMF is driven by the clump IMF, we have to ask where the clump IMF comes from. To find this we have to look further at the structures within which the protostellar clumps are found. In Figure 5.4, you can see the filamentary structure from which protostellar clumps form. More are shown in Figure 5.5, and you can see similar structures in most of the Herschel images from our own Galaxy in this and previous chapters. But these gas filaments are not unique to the sites of active star formation. The Polaris region, also shown in Figure 5.5, is free of active star formation, but is still filled by filaments of gas that have a very similar structure to those that host prestellar cores and young stars. This filamentary structure seems pretty much universal in the interstellar medium. Our current best guess at the origin of these universal filaments is that they are produced by turbulence in the interstellar medium, stirred up by the effects of supernova explosions, and radiation pressure from bright O and B type stars.

The difference between the star-forming filaments in DR21 and Aquila, and the non-star-forming filaments in Polaris would appear to be the density of the gas inside the filaments. In Polaris, there simply isn't enough gas to produce cloud cores that are dense enough to collapse under their own gravity. This may not be a permanent situation, and it may be that, over time, some of the filaments in Polaris may attract gas from other filaments in the same region to reach the threshold for star formation. This process seems to be underway in DR21, for example, where you can see sub-filaments feeding material into the denser filaments where most of the active star formation is taking place. It might also be that an external event, for example a nearby supernova explosion, could compress the filaments in Polaris, boosting their density and triggering star formation. The universal structure of these filaments, though, is what controls the size of the stars that form inside them, leading to the stellar IMF.

5.11 CONCLUSION

In this chapter we have seen that dust plays a key role in star formation process, and that star formation takes place in regions that are obscured from the view of optical astronomers. To understand what goes on inside star formation regions we have to operate at longer wavelengths, in the infrared. When we do that, we find collapsing gas clouds, young stars accreting material from disks around them, and ejecting much of that material back into space. We also find that the disks around more developed young stars match the conditions needed for the formation of planetary systems.

On larger scales, it seems that the filamentary structure of the interstellar medium is responsible for setting the mass distribution of stars. This filamentary structure is near universal in the interstellar medium. None of these insights would have been possible without telescopes that operate in the far-infrared, since the earliest stages of star formation take place in cold and dusty environments.

CHAPTER 6

Distant Heat

6.1 FORGETTING TO BREATHE

As an observational astronomer, I get to travel to various supposedly exotic locations. People get envious when I tell them I'm going to Hawaii to observe, and they don't believe me when I tell them that, to me, Hawaii is a cold windswept mountaintop, working nights.

Of course it's not just the cold and the nights that are the problem. At 14,000 feet, the Mauna Kea Observatory on the Big Island is above a third of the atmosphere. Nothing about the human body works properly at this altitude unless you've done some serious acclimatization, and even then things can go wrong. Observers sleep, when they can, at the accommodation block at 10,000 feet, but this is never restful when you're spending twelve or more hours a night on the summit. And when you're on the summit things can get quite strange.

My first observing trip provided a useful demonstration of this. You expect to get out of breath easily, and the infamous staircase at the UK Infrared Telescope (UKIRT), from the control room down to the toilets, is a case in point. It's short enough to run up, but at 14,000 feet you should never try to do this unless you like hypoxia.

But what gets to you is the unexpected.

I am sitting at a workstation, examining some of my data, when I get the feeling I'm forgetting something important. I check around me to make sure the telescope's not stopped and that we're still taking data. No trouble there. I check the weather monitor to see if the clouds are rolling in. Nothing there either, and these problems seem rather too distant, too impersonal to be what's worrying me.

I'm forgetting something important, that I do all the time by instinct or reflex, but that isn't currently working.

Breathing.

Yes, that's it!

A few deep inhalations later and I'm feeling much better, and my CO_2 levels are back to the point where they can trigger the reflex again.

It's not just small problems like forgetting to breathe that make the mountain hazardous, though. You lose IQ points the higher you go, and this can affect you in very strange ways. It's the things you least expect that catch you out, even when it comes to the simple set-up procedures. For example, we managed to waste fifteen minutes one night working out how to align an instrument on the sky purely because we couldn't remember which way was east.

But observers have to face only a limited set of problems. The people building telescopes have to face unexpected problems throughout the day as they assemble some of the most complicated constructions on the planet. And altitude doesn't help.

Hence this conversation reputedly overheard at the construction site of an observatory that shall remain nameless:

Construction worker at the summit: We have a problem with this part.

HQ at sea level: Oh yes? What is it?

CW: It's the wrong size.

HQ: The wrong size?

CW: Yes - I've cut it three times, and it's still too short!

HQ: Come down the mountain - now!

Fortunately the effects of the altitude aren't permanent. Once you're at sea level everything goes back to normal. In fact, it's a little better than normal after any time at the summit. You're suddenly reminded just how much air there is, how much oxygen there is in it, and that breathing should not be a major form of exercise.

6.2 GALAXIES - STARS AND SO MUCH MORE

Everything we have looked at so far has been in our own Galaxy, the Milky Way, but this is just one of billions of galaxies in the observable universe. If we are to understand where we came from, how our own galaxy formed, and what its future is likely to be, then we have to look further afield, and study the properties of other galaxies.

Edwin Hubble was one of the first astronomers to seriously study galaxies, and to realise that they did not lie inside the Milky Way but were similar objects to it, much further away. There are two very different classes of galaxy, as defined by their shape: elliptical galaxies, and spiral galaxies. The Milky Way is a spiral galaxy. Hubble also devised a broader classification scheme which separates spiral and elliptical galaxies into separate types, and which separates spiral galaxies into those with and without a central bar-like feature. These are known as barred spirals. As with any taxonomic scheme there are, of course, exceptions, which Hubble labelled irregulars. Figure 6.1 shows this classification scheme. The original idea was that elliptical galaxies evolved into spiral galaxies, so ellipticals are also known as 'early type' galaxies whilst spirals are known as 'late type'. This evolutionary scheme has since been found to be wrong - if anything the evolutionary processes work in the opposite

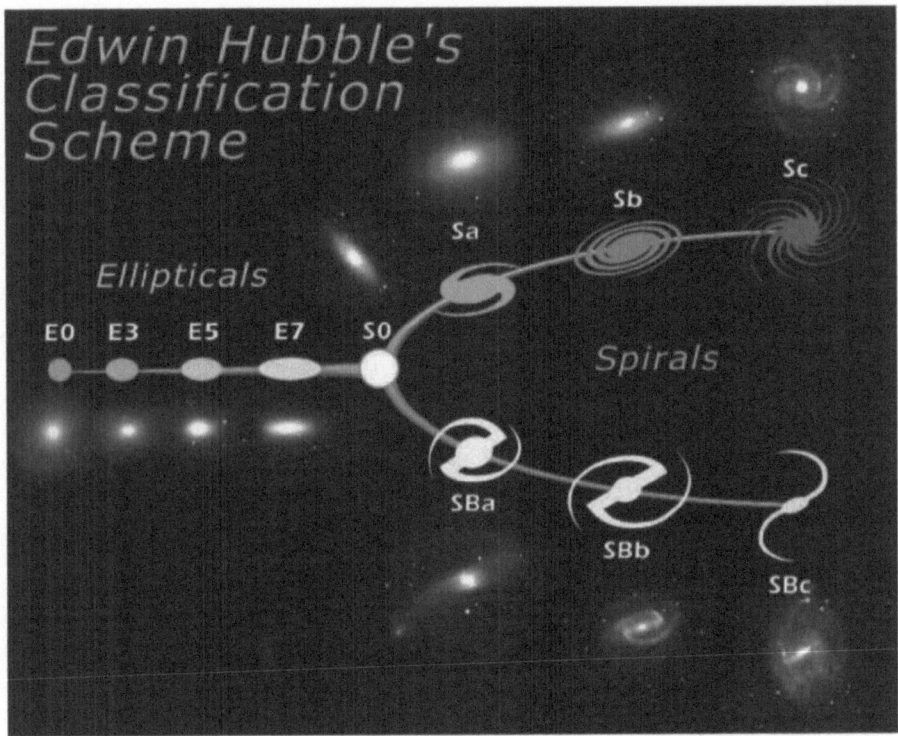

Figure 6.1   The Hubble 'tuning fork' classification scheme for galaxies.
(Courtesy of NASA.)

direction, from spiral to elliptical - but, as with so many things in astronomy,
we are now stuck with that nomenclature.

While the visible light that we see from stars in galaxies is the most obvious
thing to our eyes, galaxies are much more than just agglomerations of stars.
The majority of the mass of a galaxy is in fact made up of dark matter, the
mysterious material whose gravitational interactions have defined the large
scale structure of the universe. Each large galaxy sits inside a halo of dark
matter and, in the outer regions of both spiral and elliptical galaxies, the
motion of stars is controlled by the gravitational pull of dark matter, rather
than that of other stars. Gas and dust can also lie between the stars, making
up the galaxy's interstellar medium (ISM). The interstellar medium is the raw
material from which new stars form, as we saw in the previous chapter. There
is usually very little star formation in an elliptical galaxy, and they are largely
made up of a stellar population that is evolving passively. This simply means
that no new stars are being formed. Ellipticals are sometimes referred to as
'red and dead', their red colour coming from the lack of hot, and thus blue,
young, high-mass stars, and dead referring to their lack of star formation.

Spiral galaxies, in contrast, usually have large amounts of gas and dust, as well as active star formation, in their disks. Our own Milky Way Galaxy is a good example of this, with a star formation rate of about one solar mass of new stars being formed each year. Most of these new stars are, of course, low mass stars, given what we learnt about the stellar IMF in the last two chapters.

The gas in the interstellar medium of galaxies is similar to that found in the Milky Way. It is largely composed of hydrogen, as this is the most common element in the universe, and there will also be helium and a mixture of heavy elements that have been released into the interstellar medium by previous generations of stars. Hydrogen will be in one of three forms: molecular hydrogen, atomic hydrogen, and ionised hydrogen.

Ionised hydrogen can be detected in the optical and near-IR by its emission lines, while atomic hydrogen can be detected by its 21cm emission line in the radio. Molecular hydrogen, though, is rather more difficult to detect. It has a number of weak emission lines in the near infrared which can be excited in certain environments, but the more usual way to assess the total amount of molecular hydrogen is to look at a different molecule that acts as a tracer. The molecule usually used for this is carbon monoxide, CO, which has emission lines in the far-infrared and submillimetre. Observations at these wavelengths are thus needed to make a proper assessment of the amount and distribution of molecular gas in a galaxy.

Dust is also an important component of the interstellar medium in galaxies. Interstellar dust grains range in size from a few molecules to about 0.1 micron across. The smallest dust grains are effectively large molecules of polycyclic aromatic hydrocarbons (PAHs), which are a variety of carbon compounds made up of many interlinked hexagonal rings of carbon atoms. More exotic carbon molecules have been discovered in the interstellar medium as well, including fullerenes, the football-shaped molecules C_{60} and C_{70} named after Buckminster Fuller [63][1]. Large dust grains are made up of a carbon or silicate core onto which other molecules from the interstellar medium, including oxygen and nitrogen, have been adsorbed (see Chapter 5).

---

[1]Fullerenes were first discovered in a lab experiment trying to recreate spectroscopic features seen in the interstellar medium with laser-blasted graphite. While these molecules didn't explain all the features that we now know are due to PAHs, they did revolutionise carbon chemistry and start new industries aiming to exploit fullerenes for practical applications.

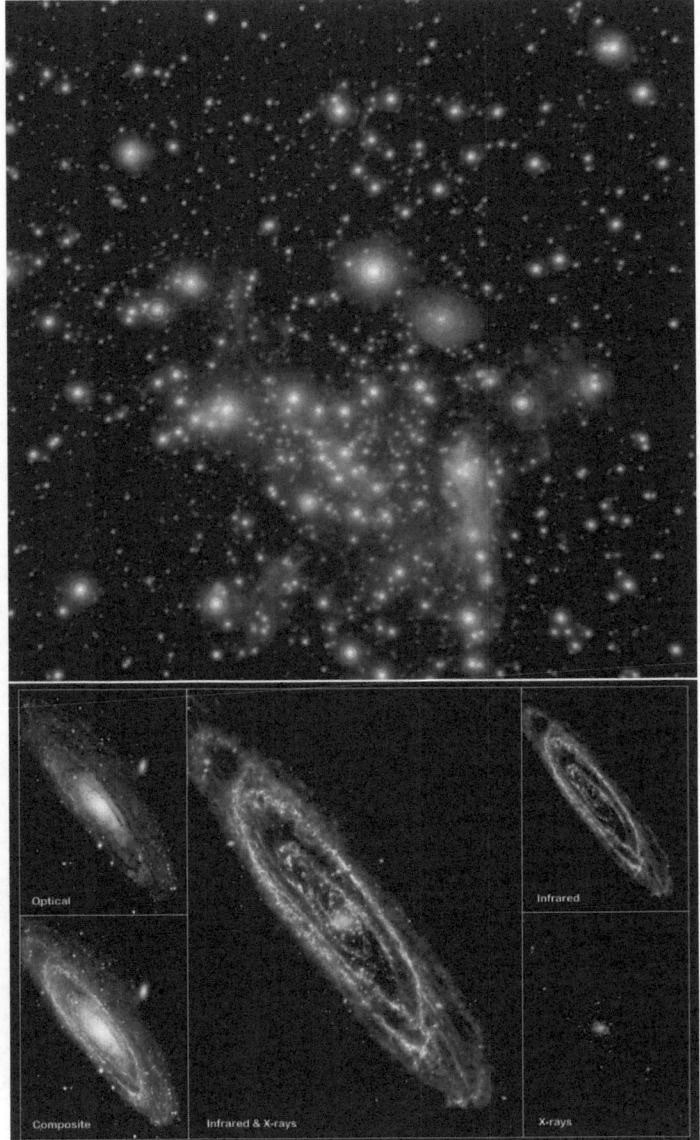

Figure 6.2 Top: A near-infrared image of stars at the centre of the Galaxy; Bottom: Our nearest neighbour, the Andromeda galaxy, seen in the optical, far-infrared and X-ray.

Top: The motion of these stars was monitored over 16 years and used to find the mass of the supermassive black hole at the centre of our Galaxy. (Courtesy of ESO). Bottom: The optical shows emission from stars, far-infrared shows emission from dust, X-ray shows the remnants left behind at the end of stellar evolution. (Courtesy of ESA.)

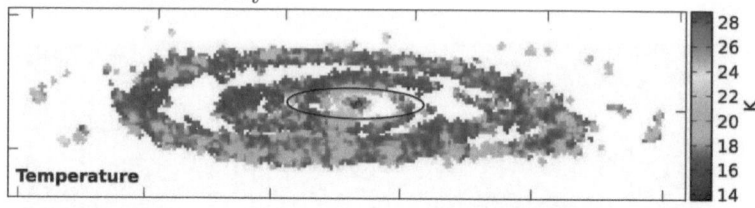

Figure 6.3 Dust temperature variation in the Andromeda galaxy.
(Courtesy of ESA.)

Figure 6.4 Top: The IRAS all-sky survey. Bottom: The Sombrero galaxy
M104, in the optical and infrared.
Top: The whole sky from IRAS showing emission at 12 microns as blue, 60
microns as green, and 100 microns as red. The red band across the centre of
the image is our own Galaxy. The blue streak across the centre to the edges
is zodiacal dust in our own Solar System. (Courtesy of NASA.) Bottom: The
Sombrero galaxy is shown in optical light (blue and green) from the HST, and
infrared light (red) from Spitzer. While this galaxy is classified as a spiral, it
has a very weak disk, but a very strong bulge. (Courtesy of NASA.)

There is one other constituent of most large galaxies that needs to be mentioned - the central supermassive back hole. Black holes are objects that have collapsed to such high densities that the velocity needed to escape from their gravitational attraction is greater than the speed of light. Normal, stellar-mass black holes are the end point of stellar evolution for some of the most massive stars. Such black holes are typically a few solar masses in size, though the remnants of supernova explosions of the most massive stars may be a few tens of solar masses. The black holes in the centres of galaxies, though, are many times this size. The black hole at the centre of our own Galaxy, for example, has a measured mass of about 4 million solar masses [64]. This was determined after a sixteen year programme monitoring the motions of stars at the very centre of our Galaxy that orbit this black hole. Optical telescopes can't do this since the dust and gas in the disk of the Milky Way obscures light from the Galaxy's centre. Near-infrared observations, which are less subject to this obscuration were used instead. Figure 6.2 shows one of the infrared images used for this study, while a movie showing the movements of these stars over sixteen years can be found via this book's website.

While you may think the black hole at the centre of our Galaxy is huge, the black holes at the centres of other galaxies can be much larger. Black holes at the centre of giant elliptical galaxies have been found with masses up to ten billion solar masses [109], over a thousand times the mass of the back hole at the centre of the Milky Way. When matter falls onto a supermassive black hole, a large amount of energy is released. This may seem counter-intuitive, since light cannot escape from a black hole, but the energy released by in-falling matter is produced outside the event horizon, beyond which nothing can escape. This happens through the formation of a black hole accretion disk, a disk of matter spiralling ever faster as it falls towards the black hole. The process is not unlike the way water spirals around the plug hole in a bath, but much faster and much bigger. Friction between the different parts of this accretion disk heat the infalling matter to many millions of degrees, and this is how the energy of an accreting black hole is generated. The high tempera-ture of the accretion disks in AGNs makes them prominent sources at X-ray wavelengths.

In most cases, galactic centre black holes in the local universe are quiescent: there isn't much matter falling onto them so they don't emit much energy. In a small number of cases, though, matter is being accreted at a much higher rate, producing what is referred to as an Active Galactic Nucleus (AGN). These objects have a wide range of properties which we will be discussing later in this chapter. While AGNs are rare locally, they were much more common at earlier stages of the universe. This is something we will look at in the next chapter.

6.3 GALAXIES ACROSS THE ELECTROMAGNETIC SPECTRUM

A galaxy's energy output as a function of wavelength, its spectral energy distribution (SED), as it is known, is dominated by different components of the galaxy at different wavelengths. The bluest stars are the most massive, with the highest power outputs, the hottest surfaces and, as we saw in Chapter 4, the shortest lives. These are responsible for emission in galaxies at ultraviolet wavelengths, and at the blue end of the optical spectrum. Since these stars are short lived, only galaxies where there is star formation underway are bright in ultraviolet and blue light. This is the case for spiral galaxies like our own Milky Way. As can be seen from Figure 6.5, which plots model SEDs for a number of different galaxy types, elliptical galaxies, lacking significant star formation, also lack the blue and ultraviolet emission that would come from young stars. At somewhat longer wavelengths, moving from the blue (short wavelength) end of the optical towards the red (longer wavelength) end, we reach wavelengths that correspond to lower temperature stellar surfaces, and thus cooler, longer-lived stars. The differences between elliptical galaxies and other types thus get smaller with increasing wavelength until in the near-infrared, where the emission is dominated by M-class stars, the SEDs are almost the same. Observations in the near-infrared can be used to determine the size of the underlying old population of stars in galaxies, with little contamination from the hot and bright, but short-lived, higher mass stellar population.

Beyond the near-infrared, we begin to see emission from material in the interstellar medium of galaxies, including gas, dust and PAHs. This material is not luminous in and of itself - unlike stars, it is not responsible for any energy generation. Instead, this material, in a variety of ways, absorbs light from stars and re-emits that energy at wavelengths different to those at which it was received. This is exactly what happens to sunlight when it reaches the earth. Rocks, sand or sunbathers on a sunny beach are drenched in light arriving from the Sun at optical wavelengths, peaking at roughly 500 nm in wavelength (0.5 microns). They absorb the energy in that light and warm up a little. As discussed in Chapter 1, any object at a temperature above absolute zero emits electromagnetic radiation with a black body spectrum peaking at a wavelength given by its temperature. For the rocks, sand and sunbathers on a beach, that temperature will be about 300 K, equivalent to 20 degrees Celsius. A black body spectrum for an object at 300 K peaks at a wavelength of about 10 microns. This is well beyond the capabilities of our eyes to detect, which is why we can't see everything around us glowing, but it can be picked up by thermal imaging cameras, as seen in Chapter 1. By absorbing light from the Sun and warming up a bit, our beach and sunbathers have, in effect, reprocessed the light of the Sun into mid-infrared radiation. In just the same way, the constituents of the interstellar medium absorb light from stars at ultraviolet, optical and near-infrared wavelengths, and reprocess it to longer wavelengths.

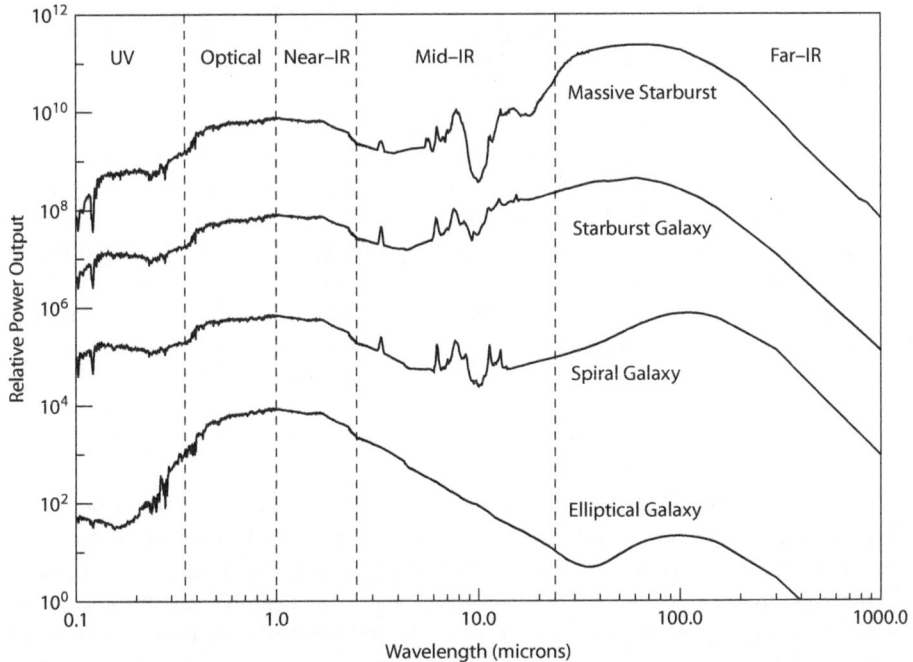

Figure 6.5 Model spectral energy distributions of several different types of galaxy.
The four different SEDs are shifted vertically so they can be compared. (Data for the model SEDs shown here are from [135]).

At the longest wavelengths, in the far-infrared, we see emission from dust at temperatures of 30 to 50 K, with emission peaking around 100 microns in wavelength. The dust responsible for this emission is made up of particles of size about 0.1 to 1 micron across. These particles may sit in the general interstellar medium between stars, or might be associated with denser regions of gas and dust responsible for star formation. The temperature of the dust will be somewhat warmer if it is closer to powerful sources of heat, such as young stars or the galactic centre black hole, and cooler if it is far from such intense heat sources and is merely warmed by light from the general population of stars in a galaxy. The total mass of cool dust in a spiral galaxy can be quite large, up to a hundred million times the mass of the Sun, but an accurate measurement of this dust mass depends on an accurate measurement of the dust temperature, and on the balance between warmer components, close to star formation regions, and cooler components.

The mid-infrared spectral energy distribution of galaxies, as can be seen from Figure 6.5, can be rather more complicated than the far-infrared. This is because several different components of the interstellar medium contribute to the shape of a galaxy's SED in the mid-infrared. First of all, there are the emission lines of different widths that can be seen as spikes upwards in Figure 6.5. These are produced by PAHs - large carbon-based molecules floating freely in the interstellar medium. Various motions of these molecules produce strong broad lines centred on wavelengths of 3.3, 6.2, 7.7, 8.6, 11.2, 12.7 and 16.4 microns, as well as some other weaker features in the mid-infrared. PAH features are usually associated with regions of star formation, and can be used to investigate the star formation underway in galaxies.

As well as the PAH emission features, there is also continuum emission in the mid-infrared, corresponding to temperatures of a few hundred to ~1000 K. This emission comes from very small dust grains, with size <0.01 microns. These grains are so small that the absorption of a single ultraviolet photon can produce a spike in their temperature to values much higher than would be expected for larger dust grains [21]. There is also a broad absorption feature in the SEDs of all actively star forming galaxies at a wavelength of about 10 microns. This corresponds to a feature in the spectrum of silicates that are a major constituent of interstellar dust.

Almost all of the features we have discussed here can be seen in the SEDs of galaxies where star formation is still under way, be that in spiral galaxies like our own, or in galaxies called starbursts where star formation is taking place much more rapidly. We will learn more about starburst galaxies later in this chapter. Elliptical galaxies, however, usually do not host any star formation, and are largely lacking an interstellar medium. Because of this, their spectral energy distribution, from the mid-infrared to the far-infrared, is largely fea-tureless. A small amount of dust, heated to fairly low temperatures of 20-30 K, can be found in them, and this produces a weak bump in their SED peaking at a wavelength of about 100 microns, but this represents an insignificant contribution to their energy output. Quite how elliptical galaxies come to have so little matter in their interstellar medium will be explored later in this chapter.

In star forming galaxies, the mid- and far-infrared can account for a significant fraction of the energy they emit. For a spiral galaxy, the mid- to far-infrared can be responsible for about 30% of the galaxy's energy emission. For the most active star forming galaxies, however, this fraction can be as high as 95%. The most actively star forming galaxies in the local universe may seem quite insignificant when observed in the optical, but can be very powerful emitters at longer, infrared wavelengths.

6.4 M31: OUR NEAREST NEIGHBOUR

The galaxy M31, found in the constellation Andromeda and therefore often referred to as the Andromeda galaxy, is the nearest large spiral to the Milky Way. It is likely that our own Galaxy would look pretty much like Andromeda if we were able to view it from the outside. The Andromeda galaxy lies about 2.5 million light years away, and can be seen as a fuzzy blob by the naked eye from a good site on a dark night. Figure 6.2 shows what this galaxy looks like at a range of wavelengths, including the optical, far-infrared and X-rays. The light from stars can be seen in the optical, while the far-infrared shows the emission from dust, and X-rays show emission from the neutron stars and supernova remnants left behind at the end point of stellar evolution. If you compare the optical and far-infrared images, you will see that far-infrared emission is strongest where there are dark regions in the optical. These are dust lanes, where plentiful dust absorbs much of the optical emission, reradiating it at longer far-infrared wavelengths. Gas is associated with this dust, so these same dust lanes are where star formation may be taking place.

By comparing the strength of the far-infrared emission in these dust lanes at different wavelengths, it is possible to measure the dust temperature [155]. This is shown in Figure 6.3. As you can see, the centre, where there are the most stars to heat the dust, has the highest temperature at about 30 K. The rings of dust that coincide with the dust lanes are at lower temperatures, but there are blobs all along these dust lanes where the temperature rises to 20 K or more. Combining the Herschel dust images with observations at other wavelengths [61] shows that star formation is taking place throughout the dust lanes in Andromeda, but the places where the dust is warmest is where stars are forming most rapidly.

Most of this star formation is caused by the normal turbulent processes of the interstellar medium (see Chapter 5), but the detailed structure of the dust in the centre of this galaxy, combined with the ring of dust lanes at larger radii seen in Figure 6.2, suggest this galaxy may have had a more violent past. These two rings taken together are thought to be the result of a collision between Andromeda and its smaller neighbouring galaxy M32 (which can be seen as the small fuzzy blob diagonally upwards and to the right of the big galaxy's centre in the optical image of Andromeda) which took place about 200 million years ago [18]. As we will see later, major collisions and mergers between galaxies can trigger large changes in their gas, dust and stellar content, as well as their shape. This more minor collision has left Andromeda largely unchanged, but M32 has lost much of its interstellar dust and gas. You can see this by comparing the optical and infrared images of M32. The smaller galaxy has far less dust, and thus far less gas, compared to its optical emission than its larger companion.

While this was just a minor incident, much larger things are in store for the Andromeda galaxy because our own Galaxy, the Milky Way, will collide with it in about four billion years [166]. This will be a much larger collision than Andromeda's encounter with M32. We will discuss what might happen as a result of this collision later in this chapter.

6.5 THE IRAS REVOLUTION

Studying our local neighbourhood in the universe is all very well, but we can only understand what is going on more broadly by observing galaxies across the whole sky. Before 1983, and the launch of the first infrared space mission, the far-infrared and much of the mid-infrared were unexplored territory. It was known that warm dust, detectable from the ground in the mid-infrared, was abundant in our own galaxy and was a good tracer of star formation, but very little was known about dust at longer wavelengths, and about its role in other galaxies. The Infrared Astronomical Satellite (IRAS) changed all that.

IRAS was the first satellite to observe the sky at mid- and far-infrared wavelengths. IRAS had only a 57 cm diameter main mirror, not much bigger than those in the largest amateur telescopes, but IRAS's mirror was not only in space, it was also cooled to liquid helium temperatures - just 4 degrees above absolute zero (i.e., -269 degrees Celsius), allowing it to make sensitive observations in the mid- and far-infrared. For a ten-month period following its launch on 25th January 1983 IRAS continuously scanned the sky, detecting sources at four different infrared wavelengths: 12, 25, 60 and 100 microns. While ground-based observations are possible at around 10 and 20 microns, IRAS was much more sensitive at these wavelengths, and was the first instrument to do significant work at 60 and 100 microns. The idea of making observations at these long wavelengths was a risk, since it wasn't clear what would be found, and getting the technology to work was a significant effort. The results, however, were spectacular. Many more sources were detected than were expected, and the scene was set for much of the work discussed in this book. IRAS discovered the first debris disk to be found, around the star Vega [10], found several comets [170] and uncovered the link between comets and meteor showers [160]. To my mind, though, it was IRAS's discovery of the 100 micron bump in the SEDs of spiral galaxies, and the even more unexpected discovery of galaxies whose luminosity is dominated by the far-infrared, that was of most importance, since it set the scene for most of the future far-infrared space missions.

The majority of the IRAS satellite's 10-month operational life was spent on an all-sky survey, mapping out the sky in its four wavebands (see Figure 6.4). What the scientists behind the mission expected to find were dust clouds in our own galaxy, and to detect a few objects outside the Milky Way, such as the nearby Andromeda galaxy. They actually found many more faint point sources, located well away from the Galactic plane, than they were expecting. The first hint that these sources were outside our own galaxy came when the IRAS satellite scanned past the Large Magellanic cloud (LMC), a small satellite galaxy to the Milky Way. At this point the IRAS team did not have an accurate measurement of where the satellite's detectors were pointing [145], so their reaction on seeing the detector scans going crazy was that IRAS had detected a large cloud of debris in Earth orbit. Eventually they realised that they were in fact seeing several unexpectedly bright dust sources in the LMC, and were able to refine the pointing of the satellite - it turned out that an on-board clock was out by several seconds, leading them to think the satellite was pointing in a different direction. Once they had the correct positions, they were able to work out that the many point sources they were finding were usually galaxies already detected in optical surveys. Examination of the optical images revealed that most of these galaxies were spirals, and comparison of the optical and far-infrared data showed that roughly a third of the energy generated by stars in these sources was being absorbed by dust and reprocessed to longer, far-infrared, wavelengths.

6.6   THE NORMAL UNIVERSE

The picture presented in Figure 6.5 is largely a product of the IRAS survey. Spiral galaxies form stars at a moderate rate of a few solar masses of new stars each year. The dust in the interstellar medium that is closest to the star formation regions absorbs more energy from the light of the new stars and heats up to temperatures of 30-40 K, while dust in other parts of the interstellar medium is heated to a lesser extent by the general light of stars in the galaxy, reaching a somewhat lower temperature of 20-30 K. This is what we see in our own Galaxy and in Andromeda, and the IRAS results show that more distant spiral galaxies behave in much the same way.

There are, however, a couple of things missing from this picture. Firstly, elliptical galaxies clearly don't work the same way, since they are largely lacking in interstellar medium, and do not have any significant ongoing star formation. Secondly, our discussion of spiral galaxies has so far been restricted to what is going on in their disks. If you look at the Hubble tuning fork diagram in Figure 6.1 you will see that spiral galaxies have not only a flat disk of stars, in which you can find gas, dust and ongoing star formation, they also possess what is known as a bulge - a spheroidal-shaped distribution of stars distributed around the centre of the galaxy, and reaching greater distances above the galactic disk than stars from the disk itself. A key feature of the

Hubble classification diagram is that spiral galaxies become more 'elliptical-like' as their bulges become a more significant part of their structure.

A good example of a galaxy with a very strong bulge and a very weak disk is the Sombrero Galaxy, shown in Figure 6.4. This demonstrates a key aspect of disk galaxy bulges: Just like elliptical galaxies, bulges are made up almost entirely of stars, and contain very little gas and dust. In fact, in a number of ways, bulges look rather like little elliptical galaxies embedded within dusty, star-forming galaxy disks. This broadens the problem, but still leaves us with the core question: How and when did elliptical galaxies lose all their gas and make all their stars?

6.7 THE ACTIVE GALAXY MENAGERIE

Another thing that became clear from the IRAS all sky survey was that active galactic nuclei (AGNs), the sources at the centre of some galaxies powered by accretion onto a supermassive black hole, have surprisingly strong emission in the 12- and 25-micron bands. This indicates the presence of hot dust in these objects. More detailed analysis of the SEDs of a variety of AGN showed the presence of dust up to temperatures of 2000 K. This is surprisingly hot. In fact, it roughly matches the temperature at which dust should sublime - i.e., boil away directly from a solid into a gas. To get so hot, and to be emitting so strongly, the dust seen in an AGN would have to be very close, in astronomical terms, to the supermassive black hole at the core of an active galactic nucleus. The realisation that dust plays a significant role in the energetics of AGNs eventually solved one of the central puzzles that had been around since the original discovery of AGNs in the middle of the 20th Century: Why do AGNs have such a wide variety of properties?

The first AGNs discovered were spotted by Carl Seyfert [152] in the 1940s. He found a number of spiral galaxies with bright point-like emission in their nuclei, the spectra of which showed emission lines suggesting gas temperatures higher than could be produced in even the most luminous stars. More than that, some of the emission lines were also very broad, suggesting that the gas responsible for them was travelling at velocities in excess of several thousand km/s. Over the next several years Seyfert found more of these sources, now generally known as Seyfert galaxies, and he classified them into two types: Seyfert 1 galaxies with broad hydrogen emission lines as well as high excitation narrow lines, and Seyfert 2 galaxies with only the narrow high excitation lines (see Figure 6.7). As more data came in, this classification became more complicated, with some galaxies having weaker broad hydrogen lines than those classified as Seyfert 1s. These were classified as Seyfert 1.5 galaxies, and soon enough people were talking about even smaller gradations in classification, with Seyfert 1.8 and 1.9 galaxies.

While the optical astronomers studying relatively local galaxies were busy with the Seyferts, radio astronomy was getting going. The first large-area radio survey to be sensitive enough to detect a few hundred sources was the

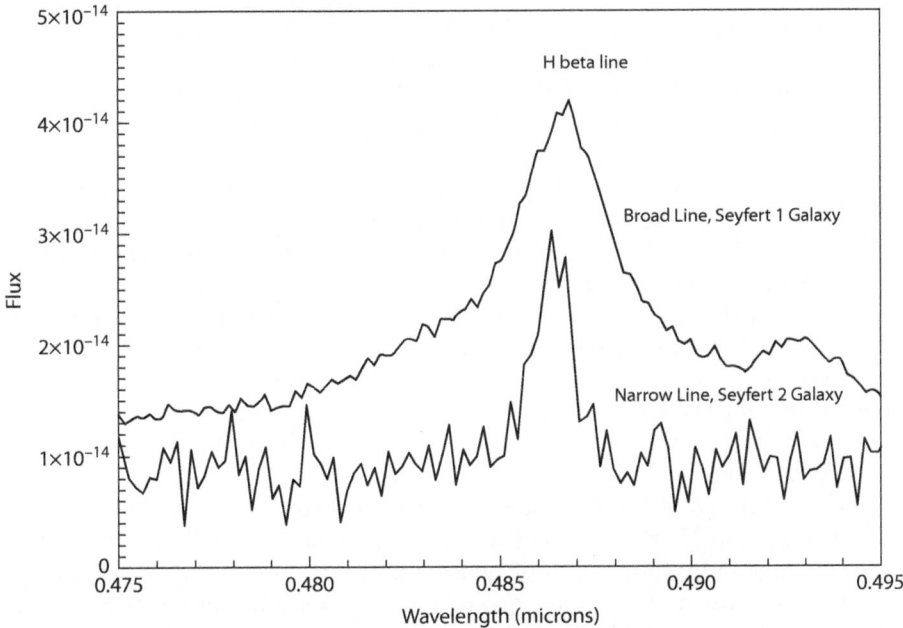

Figure 6.6 Example optical spectra of Seyfert galaxies.
These optical spectra show the distinction between broad line, Seyfert 1 type galaxies and narrow line Seyfert 2s. The line shown here is the Hβ line of hydrogen.

3C survey (the third Cambridge radio survey), completed with accurate flux measurements in the early 1960s [15]. Gradually, over the next several years, optical counterparts to some of these radio sources were identified. Some of these counterparts were clearly galaxies when viewed in the optical - these included sources like Cygnus A, which is associated with an unusual elliptical galaxy that possesses a prominent dust lane at its centre. However, a number of other 3C radio sources were found to be associated with point-like optical objects, whose optical spectra included a large number of emission lines at wavelengths that defied identification. These sources were named 'quasi-stellar radio sources' (QSOs), a title often contracted to quasar.

The mystery of QSO emission lines was finally solved when Maarten Schmidt realised that the emission lines in one of the brightest QSOs, called 3C273, could be identified if the lines were assumed to have been shifted redwards, with the wavelength of each line multiplied by the same amount. It was already known that the spectra of galaxies are systematically shifted to longer wavelengths by an amount that directly depends on their distance from us. This is known as the redshift of a galaxy, and was discovered by Edwin Hubble in 1929. It is a consequence of the expansion of the universe, and provides a useful way of measuring the distance of galaxies. We shall discuss redshift more extensively in Chapters 7 and 8. The redshift Schmidt found for 3C273 was 0.158 [149], an extremely high value at the time. When this high redshift was combined with the brightness of 3C273 it became clear that whatever was responsible for the emission in this source was very powerful indeed. We now know that quasars are powered by supermassive black holes, but it took at least two more decades before that idea was well established.

The detection of the first quasars through their radio emission was just a happy accident. Work over the decades subsequent to the discovery of 3C273 showed that only about 10% of quasars emit strongly at radio wavelengths - they are termed 'radio loud quasars' - and that the majority are in fact 'radio quiet', in the sense that they are not unusually powerful in the radio. The best way of finding quasars (and other types of AGN) is in fact to work at X-ray wavelengths, where they can be very prominent thanks to the high temperature emission of the material falling through their SMBH accretion disks.

By the end of the 1970s, astronomers studying AGN were faced with an ever increasing menagerie of different types of objects. Seyfert galaxies had been joined by radio loud and radio quiet quasars in the 1960s. Further studies of radio surveys like 3C had found a class of radio galaxies with prominent jets of radio emission, unexplainable by normal stellar processes[2]. Some of these had broad emission lines in their optical spectra, and some had narrow lines. A few radio- or X-ray-selected AGNs were found to vary very rapidly and to have featureless optical spectra, rising to shorter, bluer, wavelengths. One of

---

[2]In fact there are two classes of radio galaxies, known as FR1 and FR2, as defined by the spatial structure of their radio emission, but that isn't relevant to the current discussion.

these was sufficiently bright to have been listed as a blue, variable star in star catalogs of the 1920s, under the (incorrect) assumption that it lay in our own galaxy, and was given the name BL Lacertae (BL Lac). We now know that it lies in a galaxy at a redshift of 0.07 [124], and it has become the archetypal member of a class of AGN now known as BL Lac objects, or blazars.

This exploding zoo of different types of active galaxy was, as this brief and incomplete summary suggests, becoming unmanageable. The diversity of properties was so overwhelming that some unifying, underlying principle was needed to bring them back together. Without this, the idea that different types of AGN were all aspects of the same phenomenon might break down completely. Fortunately, detailed observations of the radio emission of blazars, arguably the most extreme AGN class of them all, provided some hints as to what that unifying principle might be. At first, though, these observations seemed to make matters worse, not better, and were initially interpreted as being a challenge to one of the central results of 20th century physics - the special theory of relativity, and the result that nothing can travel faster than the speed of light.

Observations of quasar radio jets at very high spatial resolution are able to track the movement of structures within the jets. By measuring how far these structures move along the jets over time, and taking the distance to the AGN derived from the redshift measured by optical spectroscopy, a physical velocity for these moving structures can be calculated. When this was done, a rather disturbing result was found: These blobs appeared to be moving faster than the speed of light (see Figure 6.7).

While some astronomers claimed that these apparently superluminal motions were evidence that the measured redshifts of quasars were not cosmological at all, it turns out that the answer to the apparent conflict with special relativity was already available. Some years before their observation, Martin Rees had shown that an observer looking straight down a column of material travelling towards them at close to the speed of light might perceive the material to be moving at superluminal speeds [140]. The superluminal motions weren't evidence against special relativity, they were in fact an optical illusion caused by it. The fact that these superluminal motions only occur when there is a very precise alignment between the direction of an emitted jet and the line of sight to the observer was consistent with the fact that they appear in only a very small fraction of AGNs. But what did this say about the broader population? Could it be that orientation effects, central to the perceived superluminal motions, might play a broader role in explaining the growing menagerie of different classes of AGN?

6.8   AGN UNIFICATION

An active galactic nucleus powered by the accretion of material onto a supermassive black hole can outshine an entire galaxy, but the central engine, where all the power is generated, is actually quite small. A region roughly the size of

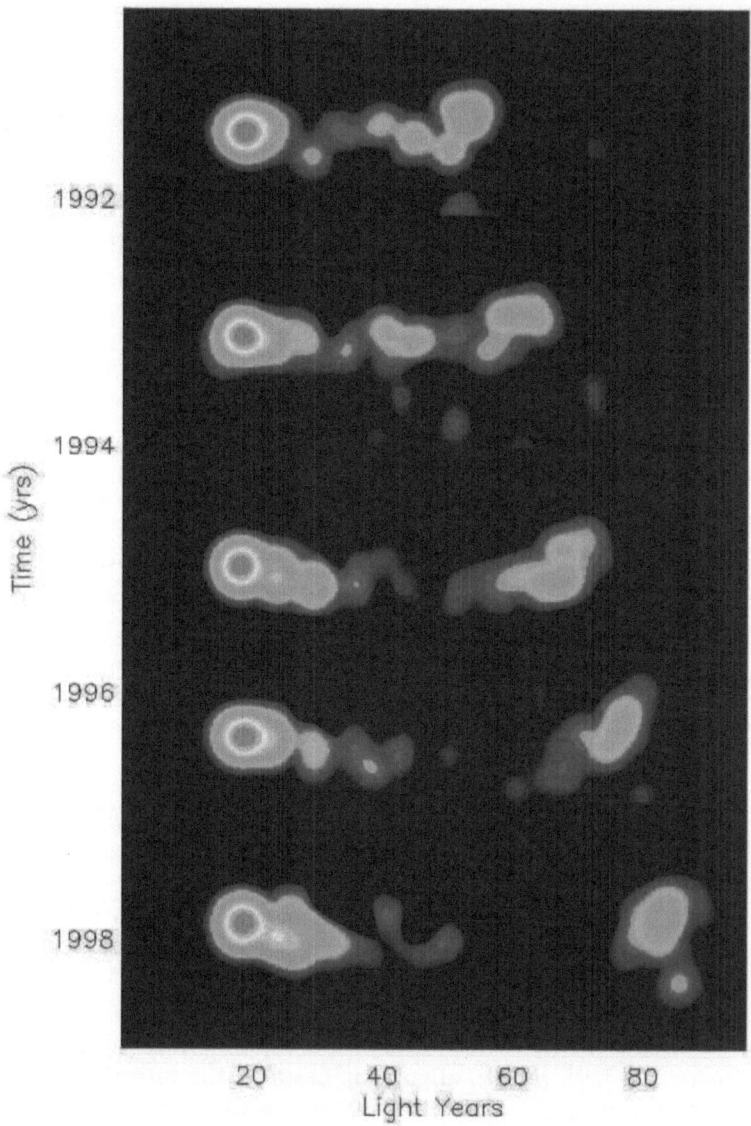

Figure 6.7 Radio images of the Blazar 3C279.

These radio images were taken over a period of six years, and appear to show structures, such as the blob at the end of this jet, moving at speeds greater than the speed of light. In this case the rightmost blob appears to move about 30 light years in just six years time, suggesting that it is moving at five times the speed of light. This apparent contradiction with special relativity is in fact the result of an optical illusion, coming from the fact that we are looking right down a jet of material that is moving very close to, but not exceeding, the speed of light [175]. (Courtesy of NRAO; https://public.nrao.edu/.)

our own Solar System would encompass the black hole itself, the accretion disk and the region of rapidly rotating gas responsible for the broad emission lines. The narrow emission lines, however, lie much further away from this central engine, at least several light years, and in more luminous quasars potentially much further. This raises the possibility that something intermediate in size between that of the broad line region and that of the narrow line region might be able to obscure the emission from the accretion disk and the broad line region, but leave the narrow emission lines unaffected. The difference between a broad-line object, such as a quasar or a Seyfert 1, and a narrow-line object, such as a Seyfert 2, would be whether we were looking through this obscuring material or not. But you can't just make the emission from the central regions of the quasar disappear. These are compact objects that can, on their own, outshine the rest of the stars in their host galaxies. If the light from the central engine is going to be absorbed, it has to end up somewhere.

This is where the IRAS result, that AGNs contain a surprising amount of hot dust, proved significant. The hot dust seen by IRAS is being heated by the central engine of the AGN, and is indicating that dust, close to the central engine, is capable of absorbing a significant amount of its emission, reprocessing it from the X-rays, ultraviolet and optical, into longer wavelength, mid-infrared emission. If dust lay in a spherical distribution all around the central engine, then all AGN would be narrow line objects, as there would be no way for us to see through to the broad-line region or the accretion disk. Instead, the dust must lie in a structure like a doughnut - something known technically as a torus - so that some lines of sight can see the central engine and the broad emission lines, but others see only narrow line emission.

The idea that the orientation of our line of sight to the structures within an AGN determine how we see it, and how we classify it, has solved many of the issues surrounding the otherwise bewildering diversity of the AGN menagerie. The picture that has emerged of the structure of an AGN, and how we then classify what we see, is shown in Figure 6.8. Viewing angles that intersect with the dust torus around the AGN see a narrow line object, those that can see past the torus to the central engine see a broad line object, while those most precisely aligned with the radio jet, in the 10% fraction of objects that emit such a jet, see superluminal motions and a BL Lac-type AGN.

This picture of the internal structure of an AGN also brings some interesting predictions, depending on the exact interaction between the dust in the obscuring torus and the emission of the central engine. If the dust acts like the interstellar dust we see in our own galaxy, then it should absorb light at shorter wavelengths more strongly than at longer wavelengths. This means that some AGN, aligned so that our view of their central engine only intersects with the edge of the obscuring dust, might reveal their central engine in the infrared, but hide it behind dust in the optical.

A good example of this is the powerful radio galaxy Centaurus A which, in the optical, appears as a an elliptical galaxy with a prominent dust lane running across its centre. Radio jets are seen to emerge from the centre of the

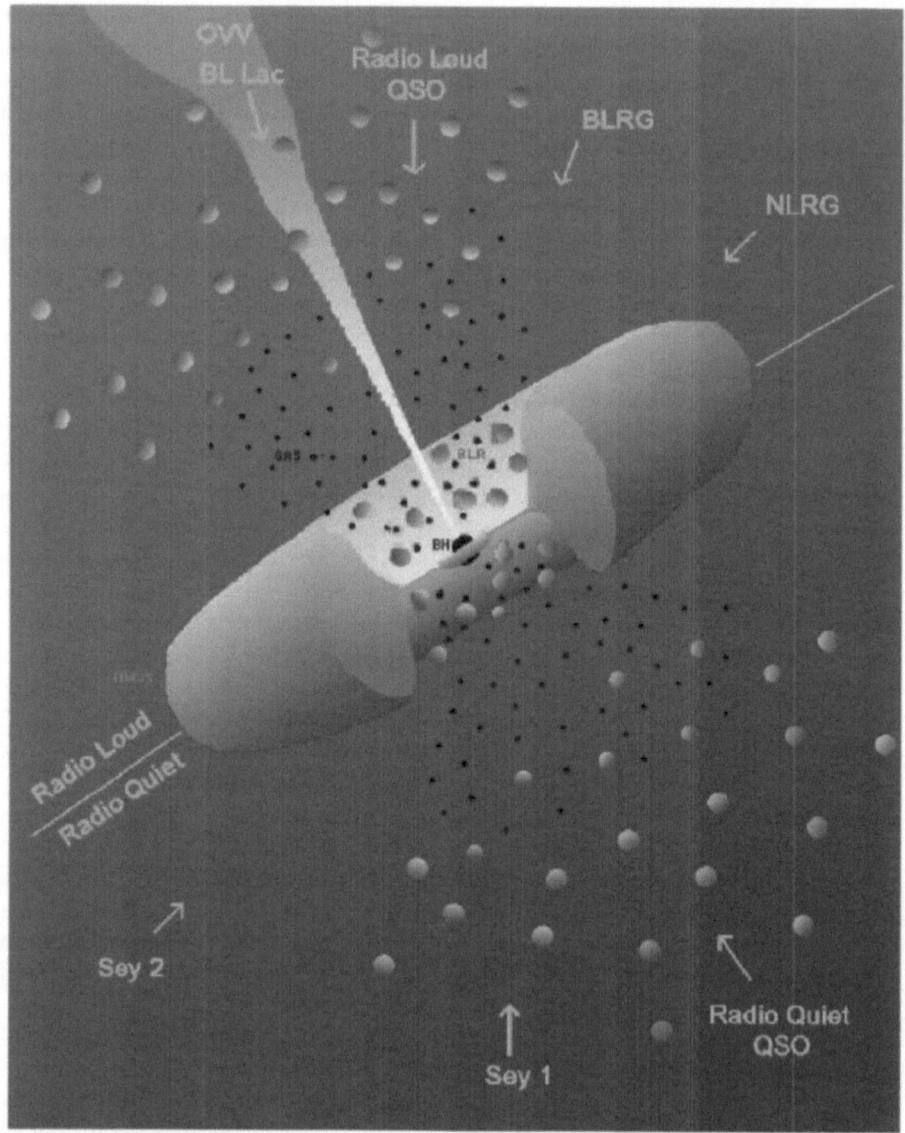

Figure 6.8 How the obscuring dust torus changes the observed appearance of an AGN.
(Copyright Astronomy Society of the Pacific, reprinted by kind permission of the authors [164].)

galaxy, presumably from an AGN, but optical light from that central engine cannot be seen because it is, presumably, absorbed by a dusty torus that surrounds it. Observations at longer wavelengths, though, can see through the obscuring dust and reveal a bright, compact source at the centre of the galaxy. This can be seen in Figure 6.10. This has been done both from space, using the Spitzer Space Telescope [88], and from the ground using a mid-infrared camera on the 8m Gemini-S telescope [138]. The latter observations, capable of reaching a higher resolution thanks to the large size of the telescope, show that the bright nucleus of Centaurus A, which outshines the rest of the galaxy at mid-infrared wavelengths, must be less than about 10 light years across. Only an accreting supermassive black hole could produce such powerful emission.

Another prediction of these so-called unified schemes is that some of the light from the central engine will be reflected by the obscuring torus rather than absorbed. This reflected light will not be very bright, but, because it has been reflected, it will be polarised, unlike the light received directly from the galaxy. For a few objects aligned in just the right way this means it might be possible to use reflected, polarised, light as a kind of periscope, able to peer into the depths of a central engine that could not be seen directly. This type of observation was first done in 1985 [5], when it was shown that the archetypical Seyfert 2 galaxy, NGC1068, had broad emission lines, the defining property of a Seyfert 1, when viewed in polarised light. A more modern example of such observations is shown in Figure 6.9.

What all these results show is that the various types of AGN are facets of the same phenomenon that appear different to us because we are viewing them from different angles. From some directions, when an AGN is being viewed through the obscuring torus of dust surrounding the central supermassive black hole, we cannot see any direct emission from the central engine. Instead, this light is absorbed by dust, which is heated to quite high temperatures, up to about 2000 K on the inner face of the dusty torus, a temperature so high that the dust is boiled away. Observations in the infrared, that can see this reprocessed emission, and using polarisation, that can, in some cases, see past the obscuration, have successfully unified the AGN menagerie.

6.9 GALAXIES AND AGN

While we only see AGN in a small fraction of galaxies, it is now generally accepted that nearly all galaxies contain a supermassive black hole at their centre, and that they all go through AGN phases during their lives. The supermassive black hole in the centre of a galaxy is ignited when a significant amount of material begins to fall onto it, triggering the formation of an accretion disk and driving all the other processes we have been discussing. As discussed above, our own galaxy has a supermassive black hole of about 4 million solar masses at its centre. If this were to ignite, it would be a fairly

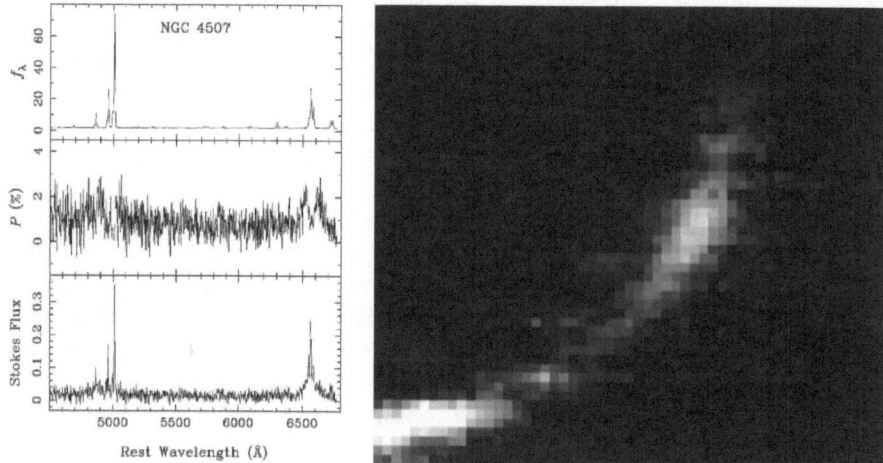

Figure 6.9 Left: The emergence of broad lines in polarised light. Right: Near infrared image of the gas cloud currently passing close to the black hole at the centre of our own galaxy.

Left: The top plot shows the unpolarised spectrum of the galaxy NGC4507. In this, the object appears as a normal Seyfert 2, with narrow emission lines. How the degree of polarisation changes with wavelength is shown in the middle figure, while the spectrum of polarised light is shown at the bottom. A much broader emission line emerges in this, which would classify the object as a Seyfert 1. The polarised spectrum is due to light being scattered off the dust torus from the otherwise obscured region around the central supermassive black hole. (Left: Reproduced by kind permission of Edward Moran. Right: Courtesy of ESO.)

weak AGN - more powerful quasars are driven by supermassive black holes that are ten or more times as massive as that in the Milky Way.

There is, in fact, a cloud of gas currently (at the time of writing) passing very close to the Milky Way's supermassive black hole [65] (see Figure 6.9). At closest approach this cloud will be just 20 light hours away from the supermassive black hole. In Solar System terms, that is a distance of 144 AU, which puts this cloud of gas closer to the black hole than the Kuiper Belt object Sedna is to the Sun. While observations of this cloud clearly show that it is being ripped apart by the gravity of the black hole, there is, so far, no indication that any material from it has fallen onto the black hole. Even if it does, the total mass of gas in the cloud is small enough that the energy released will only amount to a short-lived burp in comparison to other AGN. But this does raise the more general question: what effects do AGN have on their host galaxy, and vice versa?

For a long time it was thought that the host galaxy could have an effect on the supermassive black hole - feeding it, or starving it for example - but there was little reaction in the other direction. After all, a galaxy is a large structure, many thousands of light years across, while even the most powerful AGN central engines are much less than a light year in size. Comparisons of the properties of galaxy bulges and supermassive black holes in local, non-active galaxies, have changed that view. A series of studies, starting with a publication in 1998 by Magorrian and collaborators [104], have found a clear relationship between the central black hole mass and the mass of stars in a galaxy's bulge. This suggests that the formation of the bulge of a galaxy, a structure that can be many thousands of light years in size, and the formation and growth of the galactic centre black hole are somehow inextricably linked. Fortunately, one of IRAS's other major discoveries provides some hints as to how this might occur.

6.10   STARBURST GALAXIES

The discovery of strong mid-infrared emission from AGN, and the subsequent development of unification schemes for the various classes of active galaxies, was just one of the major successes of IRAS. Another major surprise was the discovery of galaxies forming stars much, much more rapidly than is seen in quiescent spiral galaxies like the Milky Way. These objects are forming stars so quickly that they can form a Milky Way-sized galaxy, containing a hundred billion stars, in only a few billion years - much less time than the age of the universe. They will also exhaust the gas and dust supply of a typical spiral galaxy on a timescale of a few tens of millions of years. The star formation episodes we are seeing have to be of limited duration, so these sources are known as starburst galaxies.

M82, shown in Figure 6.11, is the nearest starburst galaxy, at a distance of about 12 million light years from the Milky Way. It is a relatively small galaxy, with a total mass about one hundredth that of our own galaxy [69]. However,

M82 is forming stars at a rate of 10 solar masses of new stars each year, ten times *faster* than the Milky Way. In fact, it is forming stars so rapidly that it seems to be blowing itself apart, with a strong wind of material flowing away from the central starburst regions of the galaxy. This outflowing material is revealed by emission from ionised hydrogen and is shown in red in Figure 6.11. M82 turns out to be very bright in the far-infrared (see Figure 6.5), with roughly 90% of its total energy output emerging at these wavelengths. This is because much of the energy being emitted by the young stars born in M82's starburst is being absorbed by dust, and reradiated in the far-infrared.

M82, despite being a relatively minor galaxy in the grand scheme of things, provides several useful insights. Firstly, the wind of material being blown out by the starburst is an excellent example of how a high rate of star formation can provide feedback that acts to reduce the star formation rate, or even extinguish it completely. The starburst is blowing gas out of M82 that would otherwise lead to even more star formation, so the process is, in a sense, self-limiting. Secondly, when we try to find what might have triggered such an extreme star formation rate in such a small galaxy, we find evidence to suggest that M82 has had a close encounter with its larger neighbour, the galaxy M81. The two can be seen together in Figure 6.11 in an image taken by the WISE satellite. In this image the colours red and green represent emission in the mid-infrared, coming from warm dust, while blue represents light from stars. As you can see, despite being a much smaller galaxy than M81 - which is about 100 times more massive than M82 with roughly the same mass as the Milky Way - M82 outshines it in the mid-infrared, thanks to its prodigious star formation rate. The interaction between M81 and M82 will have caused gas in the centre of M82 to collapse, forming dense giant molecular clouds that are the site for the star formation driving its starburst. Without the interaction this gas would have remained in a stable orbit around M82's centre, forming stars at a much more sedate rate.

6.11   GALAXY MERGERS

M82 was a prominent IRAS source because it has a high star formation rate and because it is quite close to the Milky Way. Once detected by IRAS it was quite easy to identify its optical counterpart because it was already a well known object - the final Messier catalog, from which it takes its name, was published in 1781. The same could be said for many of the normal spiral galaxies that were detected by IRAS. However, many other IRAS point sources, some faint, but some very bright, were less easy to track down. When counterparts were sought by looking for optical sources at the positions of the IRAS objects, many were found to be faint anonymous blobs on the photographic plates available at the time. Others were found to have no optical identification at all. This led to a long-running series of campaigns to identify IRAS sources and to follow them up with optical imaging and spectroscopy.

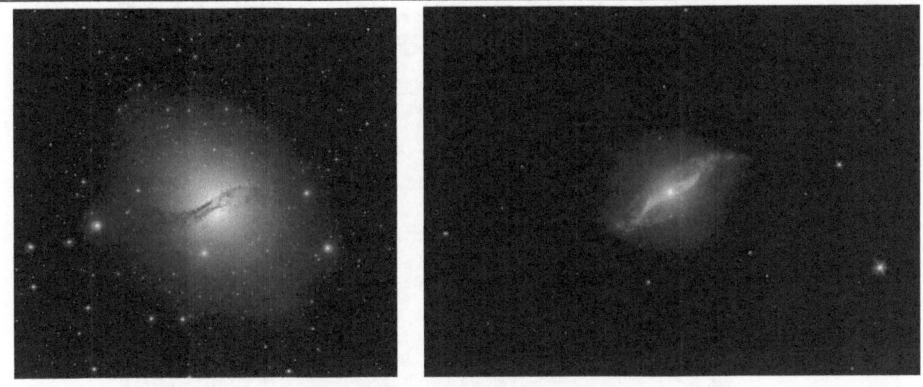

Figure 6.10 The radio galaxy Centaurus A in optical (left) and mid-infrared (right).
(Left: Courtesy of NASA. Right: Courtesy of ESO.)

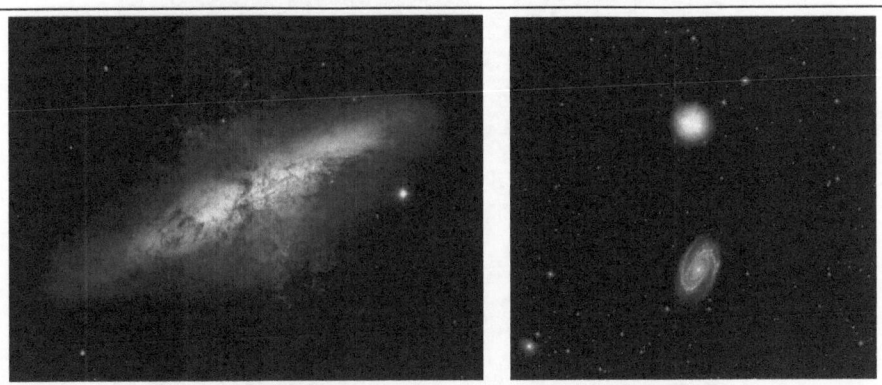

Figure 6.11 Left: M82, the nearest starburst galaxy. Right: M81 and M82 in the infrared.
Left: This optical and near-infrared image of M82 shows that it is a small disk galaxy, with plentiful dust in its disk. It also has a superwind of interstellar material blown out of the galactic disk, seen here as red emission tracing ionised hydrogen. Right: An infrared image from the WISE satellite showing M82 (upper yellow object) and its larger neighbour M81. Blue shows the emission from starlight (3.4 and 4.6 microns), while green and yellow shows emission from the mid-infrared (12 and 22 microns). M82 is much brighter than M81 in the mid-infrared thanks to its very high star formation rate. (Courtesy of NASA.)

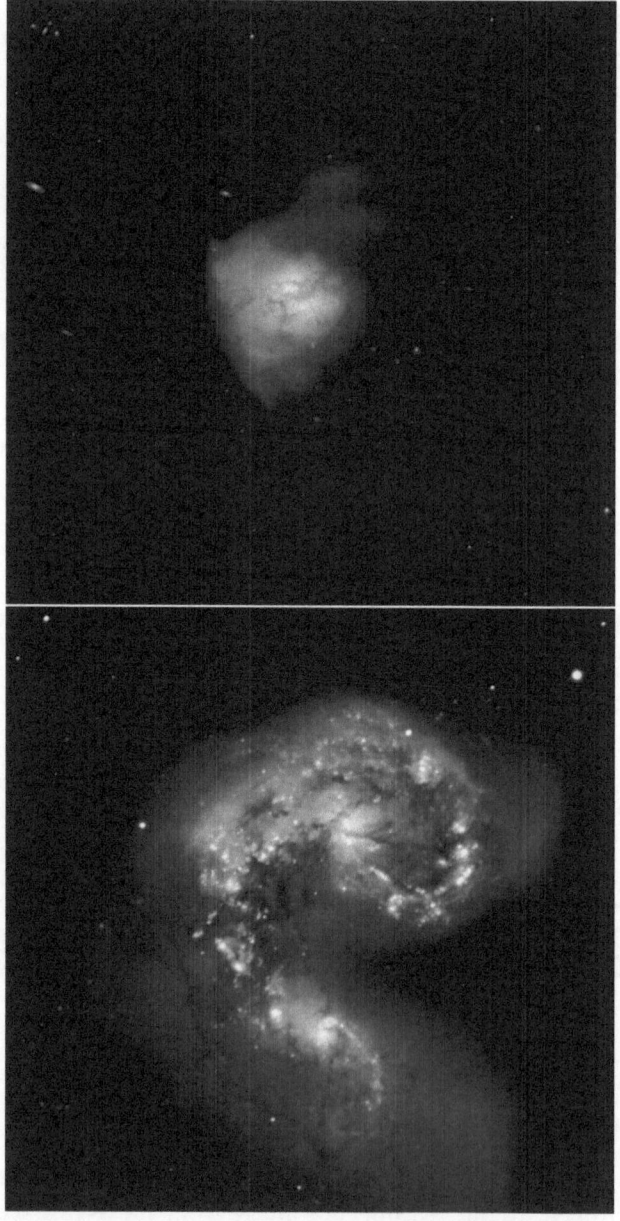

Figure 6.12   Arp220 (top) and the Antennae galaxy (bottom).
Top: An optical image showing the galaxy Arp220. Note the prominent dust
lane across the centre and the unusual shape, which is unlike either a disk
or an elliptical galaxy. (Courtesy of NASA.) Bottom: The Antennae galaxy
is a merger between two spiral galaxies which has triggered a burst of star
formation. (Courtesy of ESO.)

Much of this activity was kicked off when it was realised that one of the brightest extragalactic sources in the IRAS sky was associated with a galaxy known as Arp220. In the optical, this source isn't particularly bright. It has an unusual shape (see Figure 6.12), of which more later, which is why it was included in Arp's *Atlas of Peculiar Galaxies* [8], but at optical wavelengths it is otherwise undistinguished. In the far-infrared, though, Arp220 is an absolute monster, brighter than many galaxies much closer than its distance of roughly 250 million light years. M82 has a far-infrared luminosity equivalent to the power output of 30 billion suns; Arp220 is thirty times more luminous, emitting the power of a thousand billion suns in the far-infrared. In fact, as you can see from Figure 6.5, the far-infrared is responsible for over 95% of the energy output of a major starburst like Arp220, which is why this galaxy looks so unexceptional in the optical, despite its prodigious energy output in the far-infrared. The power source for Arp220 is thought to be a massive starburst, as in M82, but in this case the star formation rate is a hundred solar masses of new stars each year - ten times the star formation rate of M82 and a hundred times the star formation rate of our own Galaxy. All of this star formation is buried beneath thick dense clouds of dust and gas, which is why all of the light they emit in the optical and ultraviolet is absorbed by dust and reradiated in the far-infrared. This has led to these sources being called 'dust dominated' since, although dust makes up only a small fraction of their mass, they are brightest at wavelengths where dust emits.

As more and more of the IRAS galaxies were identified, more and more galaxies like Arp220, called ultraluminous infrared galaxies (ULIRGs) because of their extreme far infrared luminosity (formally a ULIRG is a galaxy with far-infrared luminosity between 10^{12}, a thousand billion, and 10^{13}, ten thousand billion, times the luminosity of the Sun), were found. At the same time, more detailed followup observations of these ULIRGs were also undertaken, trying to find out what might have caused them to be so luminous. Hints from less luminous systems, like M82 or the Antennae galaxy (see Figure 6.12), which is a merger between two galaxies as well as a starburst about ten times as luminous as M82, suggested that galaxy interactions and mergers might trigger these massive starbursts. Confirmation that nearly every ULIRG found by IRAS was a galaxy interaction or merger took many years of followup, with the case not being fully proven until the mid-1990s (something I had a hand in myself [31]), more than a decade after the IRAS satellite had died.

More detailed studies of ULIRGs and other galaxy mergers, together with computer simulations of how stars, gas and dark matter react in a galaxy collision, have shown a number of things. Firstly, galaxy mergers will drive gas and dust into the centre of a galaxy, in much the same way that the weak interaction between M81 and M82 drove gas and dust into the centre of M82. This will promote a burst of star formation. The stronger the interaction, and the more dust and gas available to drive it, the bigger the starburst will be [12]. In principle, a major merger will be able to use up all the gas in the interstellar medium of a spiral galaxy. Secondly, the effect of the galaxy merger on the existing stellar populations in colliding disk galaxies is to destroy the rather delicate structure of the galactic disks. The end result of a violent merger between two gas-rich spiral galaxies turns out to look pretty much like an elliptical galaxy, in terms of both its shape and the lack of gas and dust in its interstellar medium. The question raised earlier in this chapter, about how elliptical galaxies came to have little or no interstellar medium, seems to have been answered.

But where do AGN come into this? Alongside the triggering of a major starburst, it seems likely that galaxy mergers are capable of driving gas deep into the core of a galaxy, fuelling the supermassive black hole lurking there and triggering AGN activity [146]. At the same time, the merger-induced starburst leads to the growth of the galaxy bulge with essentially the same structure as an elliptical galaxy. The simple coincidence of these two processes is not enough to produce the strong correlations between supermassive black hole and bulge masses seen in the nearby universe. For this to happen, the galactic centre AGN must take a more active role in terminating the star formation activity in its host galaxy. This is thought to occur in a process termed 'quasar feedback', whereby the energy generated by gravitational accretion onto the galactic centre black hole in an AGN drives powerful gas outflows into the host galaxy. These disrupt the giant molecular clouds formed around the galactic centre by the galaxy merger, ending the burst of star formation.

This process can probably be seen in action in the galaxy Markarian 231. This is a galaxy in the nearby universe, lying at a distance of about 600 million light years. Markarian 231 has pretty much everything - it is a broad line AGN, a ULIRG and, as you can see in Figure 6.13, it is an interacting galaxy. What you can't see on these images, and which required far-infrared spectroscopic observations with the Herschel Space Observatory to find, is that there is a huge outflow of cool gas from the centre of this galaxy [57]. In this unusual object, we may be seeing all the processes that transform spiral galaxies into ellipticals underway at the same time: a galaxy merger, a massive burst of star formation, a new, powerful AGN lighting up at its core, and a massive outflow of material, driven by this AGN, that is even now beginning to turn off the starburst.

6.12 CONCLUSION

In this chapter we have looked at how observations across the infrared have revealed the processes that shape galaxies and the stellar populations within them, and have found how the IRAS satellite, which gave us the first all sky infrared survey, was instrumental in solving a wide range of problems behind the evolution of local galaxies. The importance of galaxy collisions and mergers was suspected before IRAS, but it took the observations of that satellite, and many years of followup studies of the sources it detected, to fully understand the linkages between galaxy mergers, star formation and AGN.

Studying these processes helps us understand how the universe came to

Figure 6.13 Markarian 231, the galaxy that has everything.
The bright, star-like source at the centre is the AGN, but this is surrounded
by a galaxy merger hosting a massive burst of star formation. (Courtesy of
NASA.)

be what we see today, but there is also a much more local reason why we might be interested in what happens during the merger of two large, gas rich spiral galaxies. That is because, in about four billion years, the Milky Way and the Andromeda Galaxy, M31, will collide and merge. That means our distant descendants, whatever they might be, will have a ringside seat to all of the processes we've been discussing here, with bursts of star formation, the ignition of galactic centre black holes as AGN, and the transformation of the two spiral galaxies into an elliptical. We're not going to be around to see it ourselves, but the view will be spectacular[3].

---

[3] A preview of what this interaction might look like has been produced by some of the people who simulate galaxy mergers, and is available via a link on this book's webpages.

CHAPTER 7

Ancient Heat

7.1   LAUNCH DAY

It's Launch Day. In just a few hours I'll find out if my scientific career is about to move into high gear, or if my future will end up in pieces on the bottom of the Atlantic Ocean.

I start early, waiting for a taxi to take me to Bush House, home of the BBC World Service. Across the UK and Europe other astronomers working on the Herschel and Planck satellites are doing the same. We're all getting media appearances out of the way before heading to wherever we can watch a live feed of the satellites' launch. After my interview, I'll be going to the Rutherford Appleton Lab near Oxford, where the UK Herschel and Planck teams are gathering with visiting dignitaries and members of the media.

I've been working on these two satellites for eight years and they've been a part of the scientific landscape since I completed my PhD in 1991. But success or failure depends on what happens today.

UK media are dominated by BBC television or, for more thoughtful coverage, BBC Radio 4. But if you want a massive audience and worldwide reach you go to the World Service. A top-rated Radio 4 appearance might reach eight million, but the World Service regularly gets audiences of over a hundred million. I try to put thoughts of this vast number of listeners out of my mind as I answer the interviewer's questions. I describe what the satellites will do and what will happen with today's launch. Then it's over and I'm ushered through the splendid art deco interior and into a taxi to Paddington Station for my journey to the Rutherford Lab.

Of the two satellites, Herschel has been around the longest, at least in terms of conceptual development. It is the last of the European Space Agency's cornerstone missions in the Horizon 2000 programme. Originally named FIRST, the Far-Infrared and submillimetre Space Telescope, it's a big expensive mission, costing about a billion Euros. All ESA countries are involved and there are contributions from NASA, Canada and China.

At the Rutherford Lab we've taken over the main lecture theatre. There's an audience of fellow scientists, administrators and the leaders of the funding agencies who paid for the satellites, as well as local politicians and journalists. The partners and children of some of the scientists have also come along so they can find out more about the project that has taken their loved ones away from home, or forced them to work late into the night.

But some people are missing. We're effectively the B-team, left at home to entertain and inform the local media and watch the launch on television. The real bigwigs are seven thousand kilometres away at Kouru, French Guiana, on the coast of South America where ESA launches its rockets. They have ringside seats for the launch and will do their bit for the visiting international media.

The closer we get to the opening of the fifty minute launch window - the period during which the spacecraft have to be launched - the more the tension rises. This isn't just a result of natural excitement. It isn't me worrying about the talk I have to give on the Planck satellite. It's because there's a ghost haunting the party, the ghost of Cluster and the launch of Ariane 501. Back in 1996, the first Ariane 5 rocket was launched carrying the Cluster satellites for the first ESA cornerstone mission. Ariane 5 was a new launch vehicle, the great hope of ESA and Arianespace, its commercial launch company. It could carry larger payloads than previous ESA rockets and was predicted by the engineers to be at least ninety-five percent reliable. I watched the 501 launch on a live video-feed in a lecture theatre not that different from today's. The rocket exploded just seconds after launch, scattering debris over the jungle and into the Atlantic. That explosion was the end of the scientific careers of several friends who had worked on Cluster. Having tied my career to the success or failure of Herschel and Planck, I'm in the same position today as they were before the 501 launch.

I give my talk, and the countdown presses closer to zero. While we have a fifty minute launch window, we aim to launch the instant it opens. Because we're launching two satellites on the same rocket, the launch itself isn't the only hazard. The satellites must get cleanly away from the launch bus before they can make their separate ways to their final destinations. Separation happens thirty minutes after launch, and only at that point can we declare it a success.

There's nothing we can do. When the Vulcain main engine and solid rocket boosters fire, we're entirely dependent on the Arianespace engineers. They have to make sure that hundreds of tonnes of high-explosive rocket fuel explodes in just the right way to get our delicate, high-precision scientific instruments, strapped on top of it all, into orbit without damage.

The final countdown begins.

'Trois, deux, une, tup!'

The Vulcain fires, plumes of steam erupt from either side of the launch pad. Then the solid boosters fire, blasting out distinctive yellow smoke, and the rocket stack is hurled upwards. Across Europe, at JPL, Caltech and online in astronomy groups all over the world, eyes are glued to screens. The roar of the rockets reaches the microphones a few seconds after the rocket clears the pad. By this point, it's already atop a vast column of flame, climbing into a clear blue sky.

Thus begins what must be the longest thirty minutes of my life. We follow the rocket every step of the way. We watch the exhaust plume as it climbs into the sky but, more important, we also follow the telemetry track from the control room, showing how the actual launch is proceeding compared to the prediction for perfect performance.

Soon the camera flicks away from the control room, now filled by our colleagues in Kouru who have seen all they can from the ground, to a view from the rocket itself as the solid fuel boosters are jettisoned - an astounding view from the very edge of space. We ignore the calls from some of the non-astronomers to open the champagne because we know there is still a long way to go.

Next is the separation of the first stage and the ignition of the second. Like every step so far, this goes like clockwork. I begin to think the launch might be going well, but desperately try to suppress these thoughts to avoid the crushing disappointment that will come if some later stage fails. The coverage is now focused on the control room, showing controllers sitting at their desks doing nothing. I am pleased to see they almost look bored.

We reach the final critical phase for Herschel, where it separates from the launch bus. We're treated to a final few frames of video taken by a camera at the back of Herschel showing its clean separation from Planck - the last external view of Planck and its Sylda 5 launch bus that anyone will see. Soon after that, it's time for Planck to cast aside the launcher itself.

Applause erupts. We start breathing again. And outside, at the reception, we hear champagne corks starting to pop.

This afternoon we can celebrate, but tomorrow we start the hard work of commissioning the instruments. It will still be months before real science can begin.

But launch, the most dangerous phase of all, has been a huge success.

7.2   GALAXY EVOLUTION

Chapter 6 looked at the state of galaxies in the local universe. Because they're close, we see them as they are now, at a time that is 13.8 billion years since the Big Bang. We saw that galaxy interactions and mergers are implicated in turning spiral galaxies into ellipticals, in triggering massive bursts of star formation and in fuelling the supermassive black holes at the centres of galaxies, triggering the emergence of active galactic nuclei (AGN). If we look at the universe today we find that about 50% of galaxies are ellipticals, though many of these are dwarf elliptical galaxies that are much smaller than the Milky Way. The fraction of elliptical galaxies is much higher where galaxies are gathered together in galaxy clusters.

If galaxy interactions and mergers turn spiral galaxies into ellipticals, then the ratio of spirals to ellipticals must have been different in the past. The elliptical galaxies we see today represent mergers of two or more galaxies that were previously gas-rich spirals. The merger converted them into an elliptical galaxy through a large burst of star formation and possibly triggered an AGN. If we were to look back in time, we would see the galaxy population changing, with more spirals and fewer ellipticals, and we might also see the star formation rate and the number of AGN increase.

Thanks to the finite speed of light we can make this exact observation. Light takes time to reach us so, in looking at sources that are a very long way away, we see those sources at an earlier stage of the universe. Looking at how the properties of galaxies change with distance, or, equivalently for our purposes here, with increasing redshift, is the study of galaxy evolution. This field has been very active over the last decade, as our observational capabilities have allowed us to identify ever-increasing numbers of galaxies and AGN at greater and greater distances.

Before we were able to study galaxy evolution in detail it was clear that the AGN population changed radically with redshift. Optical spectroscopy of quasars showed that their numbers increase rapidly as we look to higher redshifts. The number of quasars reached a maximum of about thirty times the local value at a redshift of about 2, when the universe was about 3.3 billion years old, roughly a quarter of its age today.

Quasars are much easier to detect and identify at high redshift than galaxies, which is why these results on quasar evolution were arrived at as early as the late 1980s. Similar observations of galaxies require 8 m class telescopes and concerted programmes to obtain spectroscopy of many thousands of galaxies. Studying the shape of galaxies as you look to higher redshift requires instruments capable of resolving very small scale features, since the further away you look the smaller and less distinct the characteristic features of spirals and ellipticals become. Such studies require the resolving power of the Hubble Space Telescope or large ground-based telescopes equipped with instruments that can measure and correct the distorting effects of fluctuations in the atmosphere. All this is required to study the evolution of galaxies in the optical. Studying galaxy evolution in the far-infrared is even more difficult but very necessary because, as we shall see, the optical and far-infrared stories are very different.

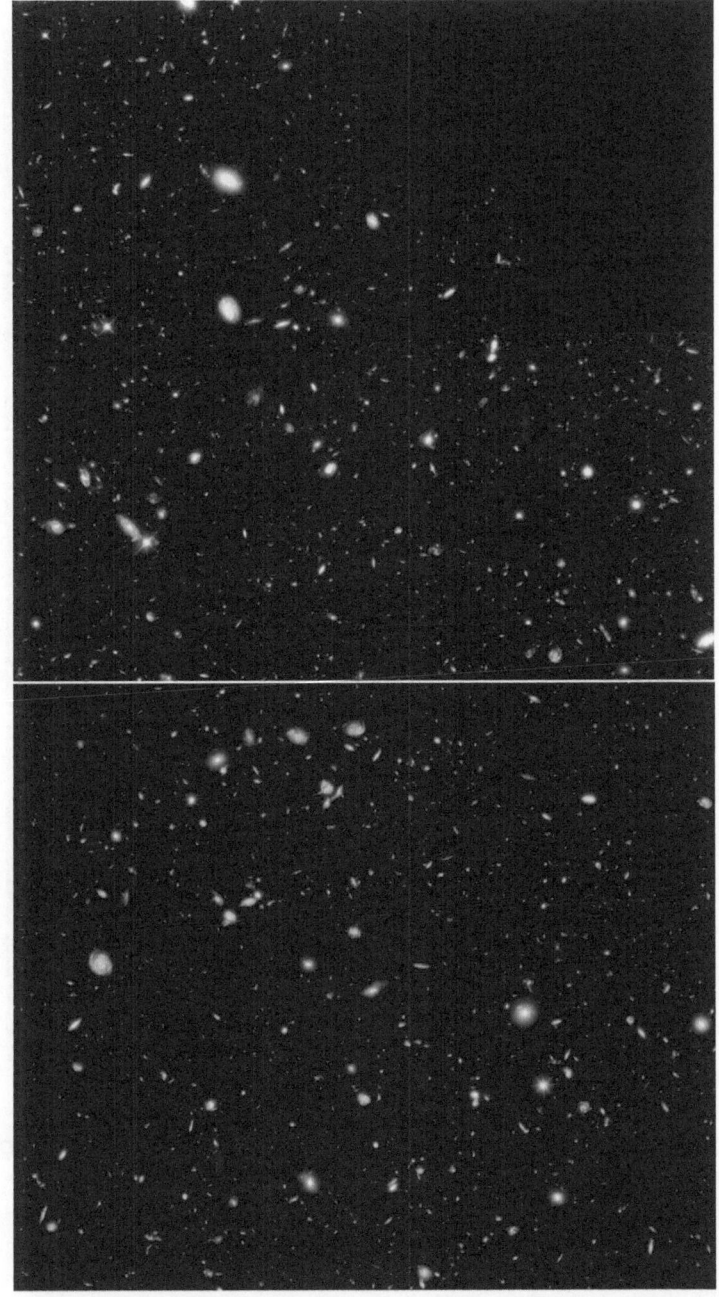

Figure 7.1 The original HDF (top) and the HXDF (bottom).
The original HDF (top) is in the optical, the HXDF (bottom) in the optical
and near-infrared, with red showing near-infrared light. Each object seen here
is a galaxy. Courtesy of NASA.

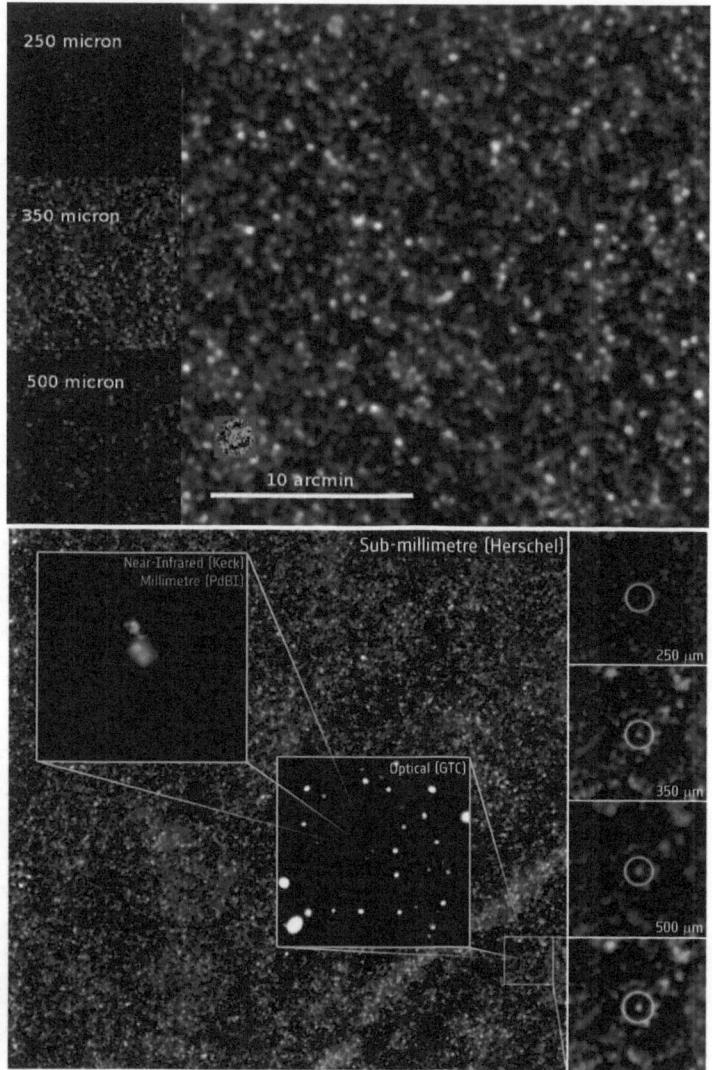

Figure 7.2 Top: HerMES Observations around the HDF; Bottom: The discovery of HFLS3.

Top: Subimages show separate SPIRE channels to left, main image shows a small part of the HerMES survey. Blue is 250 microns, green 350 microns and red 500 microns. Each blob seen here is a galaxy. Inset, to scale, is the 850 micron SCUBA map of the Hubble Deep Field. (Courtesy of ESA.) Bottom: Main image shows the region in which HFLS3 was found, colour scheme as above. At the right are images of HFLS3 at each of these wavelengths and a 'redness image' showing how it was selected. Insets show region around HFLS3 in the optical (HFLS3 is not detected) and a close-up of dust emission from the Plateau de Bure interferometer and weak near-infrared emission observed by Keck. (Courtesy of ESA.)

7.3 GALAXY EVOLUTION IN THE OPTICAL

To be able to study the evolution of galaxies at different times over the age of the universe, you have to be able to detect them at the distances corresponding to those ages. The further away an object is, the fainter it appears, so the detection of distant galaxies needs a really sensitive instrument. This meant that the study of galaxy evolution started off, and became most fully developed, in the optical, thanks to the sensitivity of 8 to 10 m ground-based telescopes and the Hubble Space Telescope (HST) in orbit.

A defining moment for galaxy evolution studies was the release of the first Hubble Deep Field observations (HDF - see Figure 7.1) in 1995 [176]. These observations represented a new paradigm for astronomy in two ways. Firstly, they made use of an unprecedented amount of time on the HST - a total of 10 full days spent with the telescope pointing continuously at a single small patch of sky. This produced the most sensitive optical image that, at the time, had ever been obtained.

Secondly, the HDF observations were made as a community service. Unlike traditional observations, the data did not belong to anybody, so nobody had a head start in analysing the data and producing results. When the data were released in early 1996, everybody could have access to it. There was a feeding frenzy. I took part in it myself, writing a paper from scratch to submission in about two weeks [32][1]. The principles set by this initial set of observations, and the tremendous scientific success they were to become, led to many similar very sensitive, openly available 'legacy' projects, including a series of ever more sensitive Hubble Deep Fields that now stretch into the near-infrared [83].

A sensitive image is not sufficient for galaxy evolution studies, even when it is the most sensitive image ever obtained. This is because extreme faintness does not immediately indicate extreme distance. A galaxy that is very faint might be very far away, but it is usually more likely that it is actually fairly nearby, and just emitting very little light. A bright thing can appear faint if it is far away, but a faint thing appears faint no matter where it is - and there are many more faint things than bright things. This is a central problem in astronomy - the night sky appears pretty much as it would if it was simply painted on the inside of a sphere, with all objects the same distance away. This was what ancient astronomers thought was going on and remained the general view until Copenicus and Kepler established the heliocentric view of the Solar System. The only objects that the naked eye can tell are moving against this celestial sphere are the planets, given that name because they are 'planetes' - Greek for 'wanderers'. Much of the study of the position of objects on the sky does not require knowledge of their distance, so it is known as spherical astronomy, since it is conducted on the celestial sphere.

But for studies of galaxy evolution we need to know the distance of sources. The usual way for extragalactic astronomers to determine distances is to obtain spectroscopy for a source, from which a redshift can be derived. Then,

---

[1]It wasn't one of my best.

thanks to the work of Edwin Hubble, who established that the recession velocity of a galaxy is directly proportional to its distance, we can obtain a distance by dividing through by the Hubble constant, now known to be 20.8 km/s per million light years (or 67.8 km/s per Mpc in astronomers' units) [129]. Spectroscopic observations take the light from a source and spread it out over a range of wavelengths. This means you need a more sensitive instrument to get spectroscopy for an object than you need to simply get an image of it. For the faint, potentially very distant, objects detected in the HDF there wasn't, and might never be, a telescope big enough to measure their spectra.

While the direct measurement of optical spectra for HDF galaxies isn't something that is going to happen soon, there are several ways of getting redshift *estimates* for these galaxies that are accurate enough for the study of galaxy evolution. These all rely at some level on examination of the observed shape of the spectral energy distribution of galaxies as revealed by their colours. A measurement of these colours was available since the HDF observations were taken in four different optical filters, covering the whole of the optical part of the spectrum. These included a filter at the edge of the ultraviolet, at 300 nm, bluewards of what the eye can perceive, towards the infrared, at a wavelength of 814 nm, at longer wavelengths than the eye can perceive, as well as one in the blue (450 nm) and one in the red (606 nm). The colours given to the objects shown in Figure 7.1 are derived from their different brightnesses in each of these four bands. More recent deep optical surveys such as the Hubble Ultradeep and Extremely Deep fields (HUDF, HXDF) [83] have added extra optical filters and, most recently, have extended these observations into the near-infrared.

The first method used for estimating the redshifts of faint galaxies in deep surveys, and the one still used to search for the most distant galaxies, is known as the dropout method. This relies not so much on the detailed properties of the source being observed, but on the presence of neutral hydrogen gas in the emitting galaxy and in the intergalactic medium along our line of sight to the source.

Neutral hydrogen will be ionised - its single electron is ejected from the atom - if it is hit by a photon with a wavelength less than 91.2 nm. So any ultraviolet light emitted in a galaxy at this or shorter wavelengths will be very strongly absorbed by even a very small amount of neutral hydrogen. The energy emitted by a galaxy as a function of wavelength drops precipitously in the ultraviolet beyond 91.2nm. Observations that match emitted wavelengths shorter than 91.2 nm will not be able to see the galaxy, while it will be much more detectable at longer wavelengths. This, all other things being equal, gives us a way of estimating the redshift of a galaxy: if a source is detected at long wavelengths but disappears when observed at shorter wavelengths, you can guess that the redshift is such that the wavelength at which it disappears - the wavelength at which it 'drops out' - corresponds to a wavelength of 91.2 nm at the redshift of the source.

But things are a bit more complicated than that, since the light from any

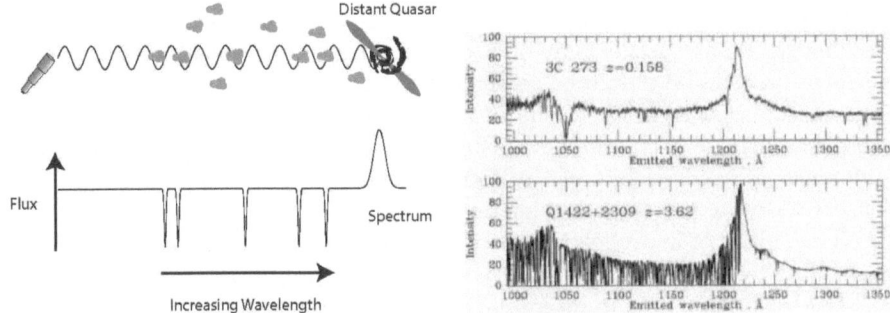

Figure 7.3 The effect of the Lyman-α forest.
Left: A cartoon showing how the Lyman-α forest is produced. Right: Comparison of nearby and high redshift quasar spectra showing the effect of the Lyman-α forest. (After an original by Edward Wright.) Right: Real observations of the effect of the Lyman-α forest, showing a local quasar, 3C273, where there is no forest absorption, compared to a higher redshift object, affected by a plethora of narrow absorption lines. The wavelengths quoted here are those at which the light is emitted, before it has been redshifted. (Figure used with permission from the Astronomy Society of the Pacific.)

distant galaxy has to travel through many millions of light years of intergalactic space to reach us, and that intergalactic space is not entirely empty. Instead, it is filled with clouds of neutral hydrogen. Neutral hydrogen not only absorbs light at wavelengths short enough to cause ionisation, it will also absorb light at wavelengths that match any of its numerous line transitions (see Chapter 1).

One particularly important line transition, which leads to very strong absorption and emission, is at a wavelength of 121.6 nm, and is called the Lyman-α line. Clouds of neutral hydrogen lying between us and a distant source will absorb light from that source at a wavelength matching that of the Lyman-α line at the redshift of the cloud. Since the cloud lies between us and the emitting source, it will be at a lower redshift than the emitter, so that Lyman-α absorption in the cloud will be at a shorter wavelength than that of Lyman-α in the emitter. Conversely, an emitting galaxy will probably have a bright Lyman-α emission line, powered by ultraviolet light coming from young stars and any AGN that might lie at its core.

A cartoon showing the process of Lyman-α absorption by clouds in the intergalactic medium can be found in Figure 7.3. The clouds are very thin, but the Lyman-α transition in hydrogen is so strong that an individual cloud will absorb most of the light of a background source at the wavelength matching that of Lyman-α at its redshift. This leads to a narrow absorption line in the spectrum of the emitting source.

Above a redshift of ∼2, these clouds are very numerous. While each cloud only absorbs at a specific wavelength, there are enough of them that they gradually eat away the light of a background source, producing a forest of absorption lines at different wavelengths. In fact these absorption lines are known collectively as the Lyman-α forest. For sources at high redshift they absorb much of the light of a source at wavelengths shorter than that of Lyman-α at the redshift of the source. This means the source drops out of observations at wavelengths shorter than this.

The clouds of hydrogen that produce the forest lines were gradually ionised over time by the ultraviolet light they are absorbing, so there are few, if any, of these clouds lurking in intergalactic space today. This can be seen in Figure 7.3, which shows spectra of the nearby quasar 3C273 and a more distant cousin at a much higher redshift. While the underlying emitted spectrum of the sources are broadly similar, you can see the effect of the Lyman-α forest in the higher redshift source in the way the spectrum is eroded away by many individual narrow Lyman-α absorption lines at wavelengths shorter than the bright emission line (which is Lyman-α emission from the quasars).

The Lyman-α forest and absorption in the emitting galaxy at wavelengths shorter than 91.2 nm mean that any distant galaxy or quasar will have little or no detectable emission at wavelengths shorter than that of Lyman-α at the redshift of emission. In a series of images taken at different wavelengths, such as those taken in the HDF, sources bright at long wavelengths that disappear completely at shorter wavelengths, are likely to be at high redshift - in fact, at a redshift that would put Lyman-α at the wavelength where the source disappears. You can see an example of this from the original HDF in Figure 7.4. This kind of colour-based approach for selecting high redshift objects is quite similar to that used for selecting cool stars in large surveys discussed in Chapter 4.

7.4 REDDER FIELDS, HIGHER REDSHIFTS

Dropout methods were originally applied to optical data from the first Hubble Deep Field and to deep surveys from ground-based optical telescopes. The ground-based surveys are not as sensitive as the HST observations but they can cover much larger areas, so the two types of instrument are very complementary. The original HDF was taken in just four optical bands. Objects that drop out of all except the longest wavelength of these bands would lie at a redshift of about five.

Pushing to higher redshift requires observations in the near-infrared, at wavelengths beyond one micron. Fortunately, the HST is up to the job, firstly with the NICMOS instrument (Near Infrared Camera and Multiobject Spectrometer), and, more recently, with the more sensitive WFC3 instrument (Wide Field Camera 3 - see Chapter 1 for details about the installation and operation of these instruments). WFC3 can observe out to wavelengths of 1.6 microns, so a galaxy visible in this filter that drops out of all shorter wavelengths would potentially be at a redshift of about 10. That corresponds to

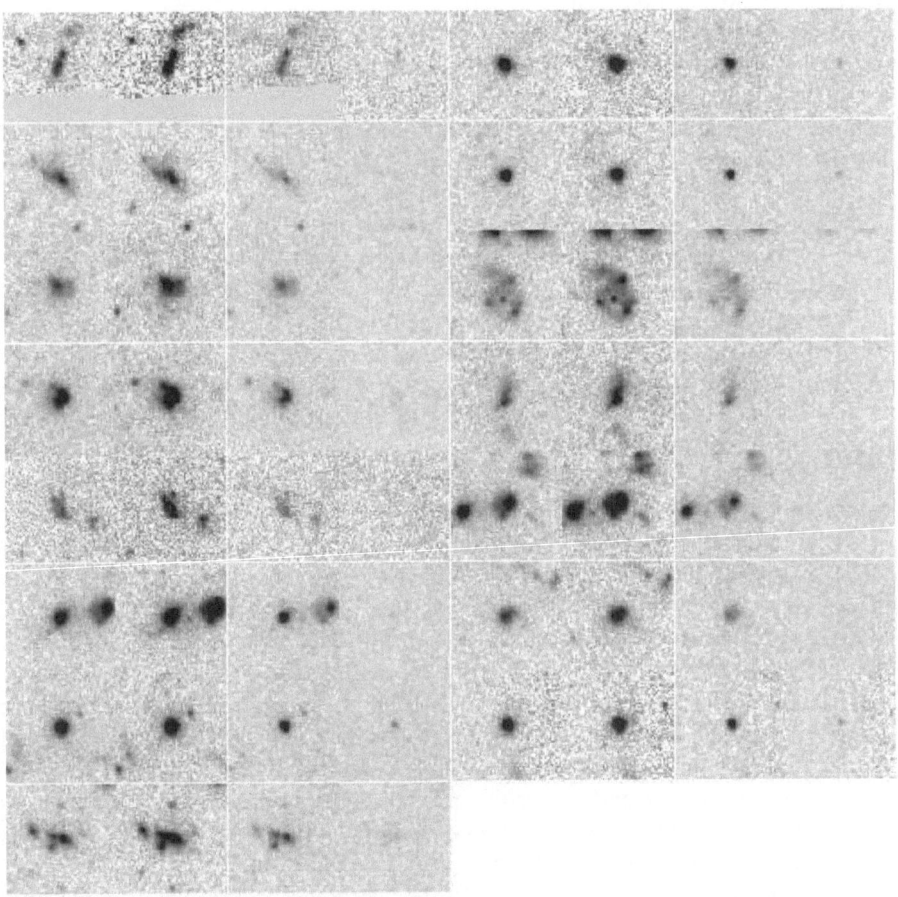

Figure 7.4 Lyman-α dropout galaxies in the HDF-South.
The images show 15 Lyman-α dropout galaxies in the Hubble Deep Field South. The images are in groups of four horizontally, with the rightmost image being at the shortest wavelength, corresponding to the Hubble F300W filter at 300 nm. All these sources are undetected or very faint in this, the shortest wavelength filter, making them candidate dropout galaxies at redshifts of about three. (Image after one in [33]).

a time when the universe was 500 million years old, only a few percent of its current age of 13.8 billion years. This is so far back in time that we really have very little idea of whether galaxies existed, what they might look like, or what kind of stars they might be home to. The discovery of galaxies at such early times would be a major result.

The descendent of the HDF using WFC3 and other updated instruments on the Hubble Space Telescope is called the Hubble Extremely Deep Field, or HXDF. It is shown in Figure 7.1 and is, currently, the deepest optical and near-infrared image ever obtained. This image is the result of the Hubble Space Telescope staring at the same small patch of sky continuously for over three weeks.

When these data were released to the public [83] there was the inevitable feeding frenzy. Astronomers from across the globe searched the catalogs and images for candidate very high redshift galaxies, with dropout selection able to find galaxies not only at redshifts around 10. Also, by looking at sources dropping out at shorter wavelengths, and because of this detected in several longer wavelength bands, galaxies at redshifts from four to eight. could also be found The advantage of these lower redshift searches is that their sources are detected in several different wavelength images, which adds confidence that they are genuine objects at the indicated redshift. Some tens of galaxies at redshifts out to eight were detected in this way [23].

Armed with a reasonable number of sources at these very great distances, it is possible to investigate the history of galaxy formation, as revealed by optical and near-infrared observations. This is done by working out the amount of energy each source is emitting - termed its luminosity - and then calculating how many sources of each luminosity are found in the volume of the survey. The quantity derived in this way is called the luminosity function, and tracking how it changes with redshift allows us to see how and when galaxies, and the stars inside them, are forming. If most galaxies form at a redshift of, say, six, but a few form at around a redshift of seven, then the luminosity function will be higher at the lower redshift since there will be more galaxies, and more stars inside them, powering them to a higher luminosity.

Figure 7.5, using data from [122], shows how the luminosity function of galaxies selected this way changes as we look from redshifts of four - corresponding to a time when the universe was about 1.5 billion years old - to a redshift of eight, when the universe was less than half that age. Between these redshifts the number of galaxies decreased dramatically, with the number of galaxies of a typical luminosity being a factor of about ten smaller at a redshift of eight than at a redshift of four. The number of high luminosity galaxies decreased by an even larger factor, perhaps by as much as 100, but these sources are so rare at a redshift of eight that hardly any are found in the small area of sky covered by the Hubble deep fields, so the uncertainties on this factor are rather large.

The total amount of star formation underway in the galaxies seen in the Hubble Deep Fields can also be calculated from these luminosity functions.

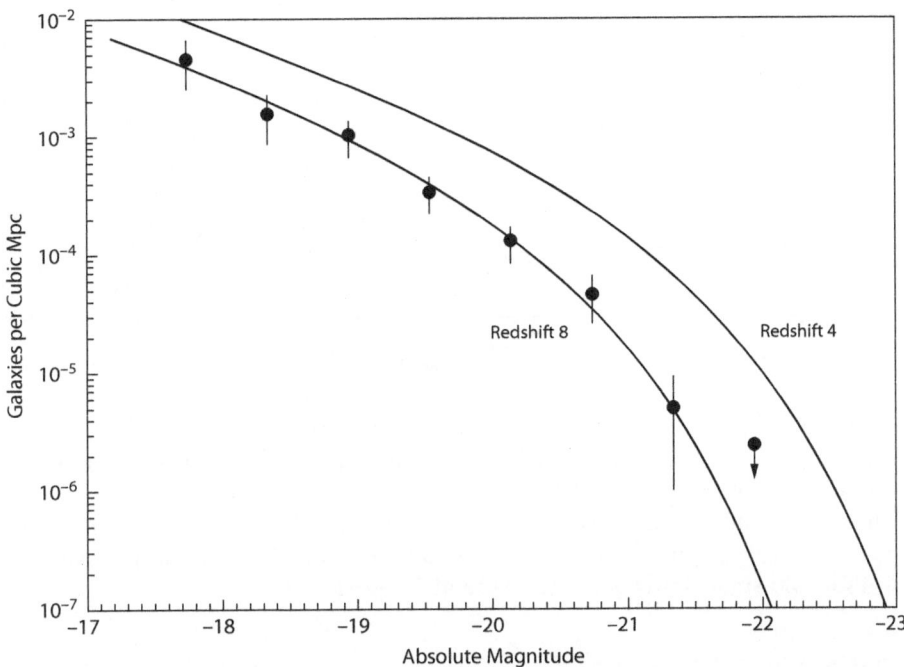

Figure 7.5 Luminosity function evolution for dropout-selected galaxies at a redshift of eight.

This shows how the number of dropout-selected galaxies changes at the highest redshifts. The most luminous galaxies are about a hundred times more common at a redshift of four than they are at a redshift of eight. (Data taken from [122]).

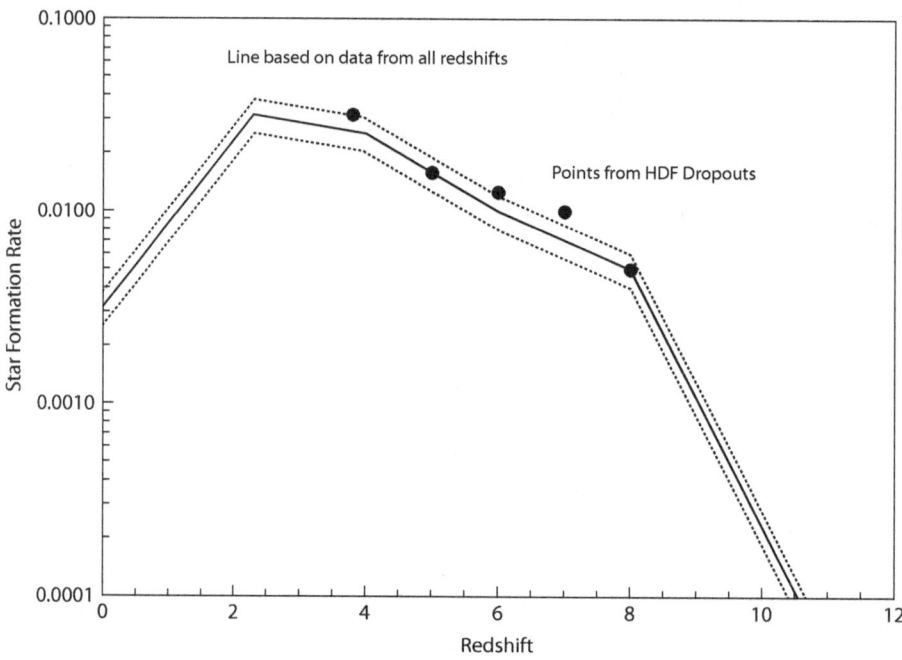

Figure 7.6 The history of star formation in the universe.
The data points at high redshift are taken from [122], while the line shows
the results of these data and a compendium of observations at lower redshifts.
The star formation rate of the universe, as determined from these observations,
peaked at a redshift of about two and then fell. The rate at the present time
is about 30 times lower than the rate at its peak.

This is shown in Figure 7.6. You can see that the overall rate of star formation
rose steadily from high redshift until it reached a peak at a redshift of about
two, corresponding to a time when the universe was about 3.3 billion years
old. The star formation rate then steadily declined over the subsequent ten
billion years. Today, the star formation rate of the universe is about 30 times
less than it was when it reached its peak. The inescapable conclusion from
these observations is that the gas that collapses to form new stars is already
running out.

7.5   THE MOST DISTANT GALAXIES KNOWN

While the Hubble Deep Fields have provided us with a good number of galaxies
at a redshift of eight, they are also capable of finding galaxies at redshifts as
great as 10, corresponding to just 500 million years after the Big Bang. Have
they detected any such objects?

The first data capable of finding such sources was released in 2009 with the Hubble Ultradeep Field, a predecessor to the deeper HXDF data shown in Figure 7.1. The initial indications were good, with several candidates found that appeared to be detected in just the longest wavelength, 1.6 micron, near-infrared band. But as further work was done, this list, small to begin with, was gradually whittled down to just one candidate that survived all the tests. A paper on this source, potentially at a redshift of 10, was published in the prestigious scientific journal Nature [22].

The story doesn't end there, though. More observations have since been collected, producing the even more sensitive HXDF, and the data on the initial redshift 10 candidate has been studied intently by other groups. The problem with a source like this is that it is only detected in a single waveband, and it is so faint that it is very difficult to obtain a spectrum for the source. A good spectrum would provide an unambiguous measurement of the redshift, but without that, there are other possible explanations for its appearance. The source might not be real, for example, but be an unfortunate statistical conspiracy between different sorts of noise that produce the appearance of a source. The deeper HXDF data put paid to that hypothesis - the source is definitely real. However, there are also more subtle alternative possibilities. The source is real, but, rather than lying at a redshift of 10, could it be closer to us? Could its appearance in a single imaging band be due to an unusually bright emission line at an appropriate wavelength in a lower redshift and otherwise very faint galaxy? Could it appear very red not through its high redshift, but as a result of dust obscuration reddening the spectrum of a more nearby object?

These and other ideas have all been checked for this particular source over the last several years, and it now appears that it might not be as distant as it was once thought. The team that originally discovered this object obtained some very sensitive spectroscopic observations with the Hubble Space Telescope that suggest the presence of an emission line from oxygen at the much lower redshift of 2.19 [25]. The brightness of this emission line compared to that of stellar light in the same object is very surprising, but the same study finds other sources with similarly strong emission lines. The unfortunate conclusion is that the first claimed redshift 10 galaxy is probably a much more nearby object.

Further studies, both in the HXDF [123, 53] and elsewhere, have produced other very high redshift candidates, and some of them may well be real. However, the possibility that they are lower redshift sources that are being picked up because of bright emission lines makes it difficult to be absolutely sure about any of them. What is needed is a sensitive near-infrared spectrometer in space, more sensitive than the Hubble Space Telescope can provide. Fortunately such an instrument is planned as part of the James Webb Space Telescope (JWST), due for launch towards the end of this decade. We will say more about this project in Chapter 9.

7.6 THE COSMIC INFRARED BACKGROUND

The picture of high redshift galaxy evolution that has emerged from optical and near-infrared studies over the two decades since the first Hubble Deep Field is one of relatively quiet change. The sort of objects responsible for the star formation history of the universe seen in Figure 7.6 are systems which change slowly, grow stars consistently over a long period of time, and which have little or no excitement in their lives. The kind of fireworks we see in the local universe in objects like Arp220 are not to be found at high redshift in this picture.

But it turns out that this picture is incomplete.

Hints that this is the case, and that far-infrared luminous, starbursting galaxies like Arp220 have a far bigger role in the history of galaxy evolution than suggested by optical studies, first came from IRAS. Studies of the most luminous galaxies detected by this satellite found that their numbers were increasing rapidly with redshift, indicating that they might be much more common at the redshifts where most star formation took place [147]. Of course this is easy to say now, with the benefit of hindsight.

The true significance of dusty starburst galaxies in the history of the universe only became clear in the late 1990s, when the results from two instruments on the Cosmic Background Explorer (COBE) satellite were announced. COBE's main job was to look for small fluctuations in the cosmic microwave background (CMB). This is the dull glow, at a temperature of 2.73 K, left over by the Big Bang. The CMB, the Big Bang, and what it can tell us about the origin of the universe will be the subject of the next chapter.

The two instruments on COBE of interest to us here are the Diffuse Infrared Background Experiment (DIRBE) and the Far-InfraRed Absolute Spectrophotometer (FIRAS). The goal of these two instruments, in the current context, was to search for any other diffuse backgrounds, other than the CMB, that might exist. DIRBE did this by observing the sky with a small, 19 cm, telescope in ten different bands from the near- to far-infrared. FIRAS, in contrast, measured the spectrum of the sky from the far-infrared into the radio, taking an accurate measurement of the amount of energy received at about 40 different wavelengths. It's main job was to determine how closely the CMB matched a black body spectrum[2], but, once that was determined, the CMB could be removed and any other backgrounds could be examined.

Two completely independent teams analysed the data from these two instruments in search of any background emission that might appear at wavelengths between those of the CMB and those in the optical and near-infrared produced by the light from galaxies like those seen in the Hubble Deep Fields. In 1996, a French team published the surprising result that the FIRAS data revealed a far-infrared background, whose spectrum peaked at a wavelength of about 200 microns [137]. A lot of careful analysis was needed to remove

---

[2]So closely, it turned out, that the CMB was a better black body than the calibrator on the instrument.

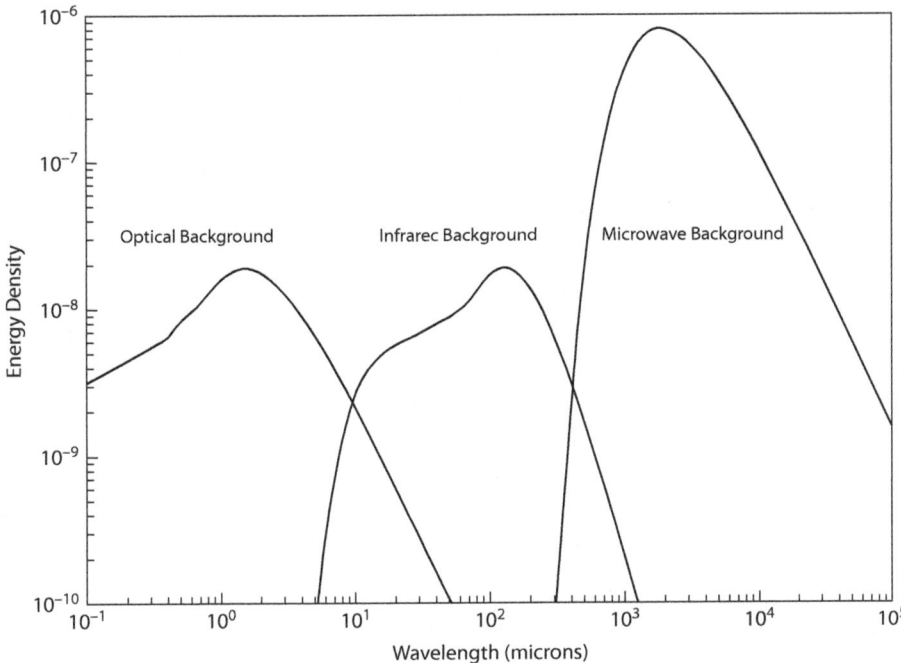

Figure 7.7 The cosmic optical, infrared and microwave backgrounds compared.
The energy density of the CMB is the highest, but the optical (COB) and infrared (CIB) backgrounds contain similar amounts of energy. (Modified from an original figure by H. Dole et al. *The cosmic infrared background resolved by Spitzer. Contributions of mid-infrared galaxies to the far-infrared background.* A&A, 451:417429, May 2006 [46]).

other contaminating sources from this background, and the results remained somewhat noisy, so the paper only described this detection as 'tentative'. Confirmation, though, of the existence of a cosmic infrared background (CIB) arrived two years later with the results of the DIRBE instrument [58]. The two data sets matched perfectly. Even more surprising was the amount of energy in the CIB. When this is compared to the amount of energy emitted by all the distant optical and near-infrared galaxies found in deep surveys such as the Hubble Deep Fields, which can be thought of as the cosmic optical background (COB), the two are found to be almost identical. This means that the sources that power the CIB have, over the history of the universe, generated as much energy as all the stars we can see in all the galaxies observed by our optical telescopes. This can be seen diagrammatically in Figure 7.7, which also shows that both of these backgrounds still contain much less energy than the CMB - not that surprising since the CMB comes from the formation of the universe.

What are the sources that power the CIB? At the time of its discovery this was unclear, since there were no far-infrared instruments that were sensitive enough to probe the details of its structure. The suspicion, though, was that the CIB would, like the COB, be made up from contributions of many individual galaxies emitting at the appropriate wavelengths. Far-infrared light, as we have already seen, is the result of dust absorbing light from some source of energy, usually stars and star formation, and being heated to temperatures of about 50 K. The discovery that the CIB and COB each encompass comparable amounts of energy meant that half of the energy generation in the history of the universe is obscured by absorbing dust. Optical surveys, even the most sensitive ones available, will miss this energy, so the history of galaxy formation revealed by the Hubble Deep Fields is incomplete - it is telling only half of the story.

The immediate assumption of many astronomers was that the CIB and COB are generated by the same types of galaxies, and that everything they had discovered in the optical could account for the CIB just by doubling all the numbers, on the basis that half of the emission of their sources was absorbed by dust. When the first results from the new generation of instruments on ground-based telescopes capable of saying something about the sources responsible for the CIB were announced, something rather different was found.

7.7   SCUBA AND SUBMILLIMETRE GALAXIES

In the late 1990s a brand new, and much anticipated, instrument arrived at the James Clerk Maxwell telescope in Hawaii. It was called SCUBA - the Submillimetre Common User Bolometer Array [77] - and promised to revolutionise our knowledge of dusty objects. This is because it operated in the submillimetre - at wavelengths just shorter than 1 mm, where the atmosphere is reasonably transparent, but where the long wavelength tail of far-infrared emission from dust is still detectable - and was several hundred times more powerful than previous instruments at these wavelengths. It managed this huge stride in capability in two ways. Firstly, by having multiple bolometer detectors (see Chapter 2 for a description of how bolometers work) - 128 in total - compared to the single detector in the previous submillimetre instrument at JCMT known as UKT-14. Secondly, each of the bolometer detectors inside SCUBA was several times as sensitive as the single detector in UKT-14. Multiply those factors together, and the stage was set for something rather special.

The Hubble Deep Field was an obvious first target for deep extragalactic observations with SCUBA. In early 1998, during the first months of SCUBA operation, a total of fifty hours of observing time was spent staring at the Hubble Deep Field [81], in what amounted to one of the single longest exposures in the nearly 9 year operational life of the instrument. The results of these observations can be seen in Figure 7.8, which can be compared to the optical view of this same field in Figure 7.1.

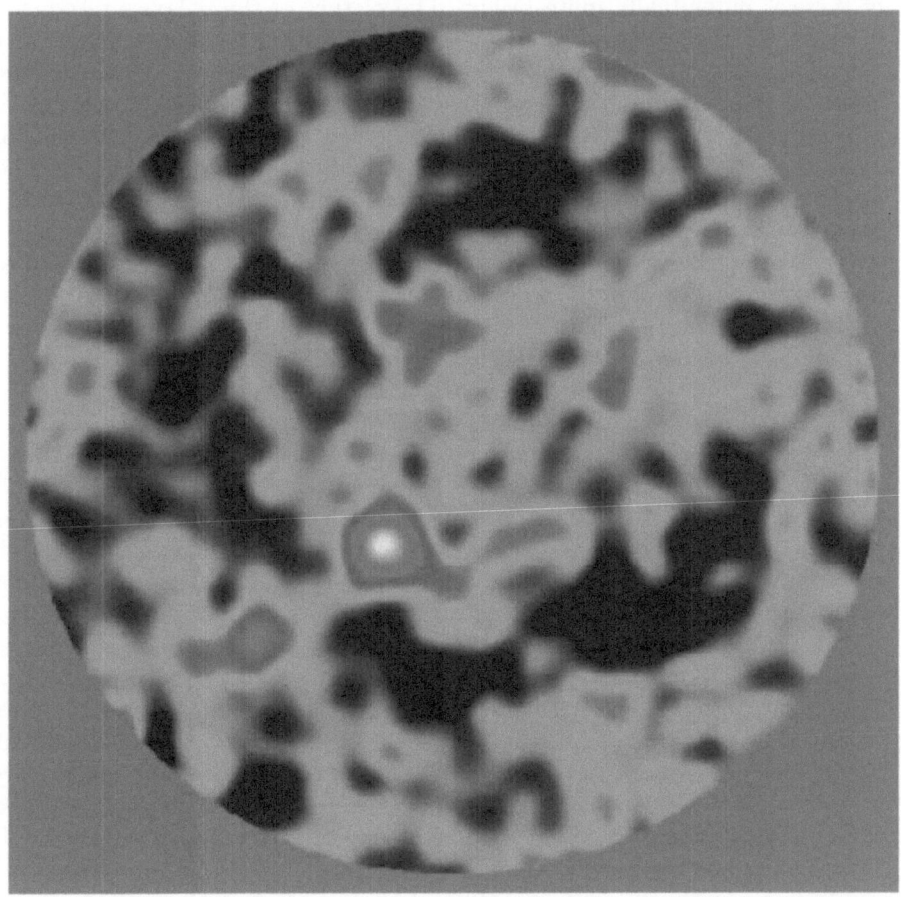

Figure 7.8 The SCUBA image of the Hubble Deep Field at a wavelength of 850 microns.
HDF850.1 is the brightest source in this image, visible as the white blob just to the lower left of the centre of the image. (Taken from [77], courtesy of the Royal Observatory Edinburgh.)

These two views of the universe, one looking at the sources that make up the COB, the other looking at the sources that make up the CIB, are very different, as you can see. While the optical image of this field reveals about two thousand relatively faint galaxies, the SCUBA image, showing emission at a wavelength of 850 microns, finds only a few sources - five in total - but each of these is individually quite bright. The picture of the galaxies responsible for the CIB that emerges from these observations is strikingly different to what has been found from the deep Hubble observations of the galaxies that make up the optical background. While the latter evolve slowly and quietly, forming stars at a moderate rate of one or two solar masses per year, the CIB galaxies, as revealed by SCUBA, are individually much more luminous. They are forming stars a hundred times faster than the galaxies behind the optical background. If we compare these distant galaxy populations to the kind of galaxies we see locally, the galaxies contributing to the optical background are like our own Milky Way Galaxy, while the CIB sources revealed by SCUBA are similar to Arp220 and other starbursts. Like the local starbursts, the phase of rapid star formation we are seeing in the SCUBA objects cannot last very long as they will use up all the gas available to them.

After the observations of the HDF with SCUBA, a new industry of surveys at submillimetre wavelengths got underway. Over the lifetime of SCUBA, the first five submillimetre galaxies (SMGs) detected in the HDF were joined by a few hundred others that were uncovered in surveys that, when they are all added together, amounted to a bit less than one square degree on the sky (the whole sky amounts to a little over 40000 square degrees). The most ambitious of these surveys, known as SHADES, short for the SCUBA Half A Degree Extragalactic Survey [114], aimed to cover half a square degree, reaching sensitivities not quite as good as the SCUBA HDF observations. These observations took place over a seven year period [36] and, as a result of technical failures that led to the retirement of SCUBA in late 2005, the full half degree area was never completed. Nevertheless, SHADES discovered over a hundred new SMGs which was, at the time, a significant step forward from the handfulls of sources found in surveys like the HDF and the Canada-UK Deep Submillimetre Survey (CUDSS) [50].

Analysis of these surveys confirmed that the population of far-infrared luminous galaxies evolved very rapidly with redshift, with this type of source being up to a hundred times more common at a redshift of two, about ten billion years ago, than they are in the local universe. The cause of this rapid evolution, and the origin of the spectacular starbursts that power both these individual objects and the CIB as a whole, is still unclear. One possibility is that galaxy interactions and mergers were much more common at earlier stages of the universe, as the galaxies we see today assembled themselves through gravitational attraction from the smaller galaxies that formed earlier in the universe. Another is that these galaxies are growing through the gravitational accretion of cool hydrogen gas from their surroundings, and that this infalling gas is fuelling bursts of star formation.

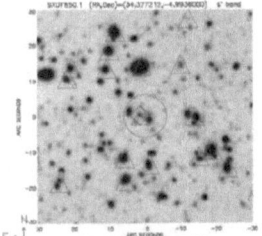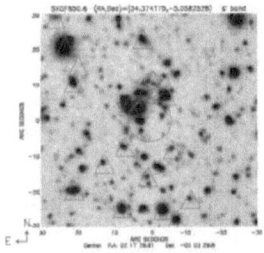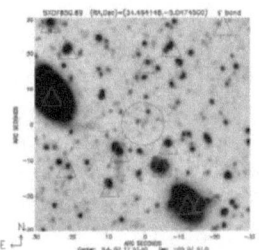

Figure 7.9   The identification of SMGs from the SHADES survey.
These show optical images (greyscale), the positions from SCUBA (circle),
the positions of sources detected in the near- and mid-infrared (triangles)
and radio (diamonds) for three SHADES sources. In the leftmost case a clear
identification is indicated by the radio data. For the middle source there are
three plausible identifications in the radio, while on the right, there is no clear
identification inside the SCUBA error circle [34].

7.8   SUBMILLIMETRE GALAXIES AT OTHER WAVELENGTHS

To test the different ideas for the origin of the vast bursts of star formation
that power SMGs we have to study them with the full range of observational
facilities available to astronomers. Before that can be done, we have to identify
their counterparts at other wavelengths. This turns out to be harder than you
might think. The precision with which the position of sources can be measured
with SCUBA isn't particularly good. On the basis of these positions alone,
there might be several tens of possible optical counterparts for any given source
in the SCUBA HDF, or in any of the other SCUBA surveys. To home in on the
correct identification, we have to rely on the fact that sources that are bright in
the submillimetre and far-infrared, under certain assumptions, are likely to be
bright at other wavelengths, including the mid-infrared and the radio. We can
then use observations at these other wavelengths to measure the position of an
SMG more precisely, and so track down their optical counterparts. An example
of this process is shown in Figure 7.9. These are optical images centred on the
positions of sources from the SHADES survey [34]. The positional uncertainty
from SCUBA is given by the circle at the centre of the image. If you look at
the leftmost part of the figure, you can see there are plenty of optical sources
that lie inside this circle, and so plenty of plausible optical identifications.
But there is no way of telling which of these is the correct identification using
the SCUBA position and the optical data alone. The triangles indicate the
positions of sources found in the SWIRE near- and mid-infrared survey using
the Spitzer space telescope [101], while the diamonds indicate the positions of
sources detected in the radio using the Very Large Array [86]. With this extra
data you can clearly see that the correct identification for the leftmost SMG
is the faint optical galaxy indicated by the diamond.

Once an optical identification has been obtained, a wide range of followup observations become possible. The most important of these is the ability to measure a redshift using optical spectroscopy. In 2005, after a heroic effort at cross identification and followup optical spectroscopy, the astronomer Scott Chapman and collaborators [29] showed that SMGs identified in this way lay at redshifts of around 2.3. Once the distance of these sources were known, many of their physical properties, such as their size and power output, could be calculated.

However, the identification process used by Chapman and others has its flaws. As you can see in Figure 7.9, while this process works well in many cases, in some, such as the source in the middle, the results can be ambiguous, while in others, such as the source on the right, the process fails completely. At redshifts above 3, the brightness of the sources in the radio or near- and mid-infrared are expected to fall off. More distant sources, which will also be the most powerful and most interesting, are likely to be missed by this particular approach.

7.9 THE HDF850.1 STORY

The brightest submillimetre source found in the Hubble Deep Field, given the prosaic name of HDF850.1, is a great illustration of the problems of identifying the most distant submillimetre galaxies. This source was originally found in the HDF SCUBA observations in 1998 [81]. As the brightest source found, it might be expected to be the brightest at other wavelengths as well, but this turned out not to be the case. Over subsequent years there were observations at a range of different observatories and wavelengths. A better position was obtained from millimetre interferometers [49], which led to claims that it was associated with an optical source at a redshift of about 1.7, but there remained a significant disagreement between the millimetre and optical position. More and more sensitive observations at a range of wavelengths were acquired, from X-ray to radio, but still no convincing identification was made. There were even suggestions that HDF850.1, and, by implication, many of the other unidentified SMGs, weren't galaxies at all, but were cold dusty gas clouds in our own galaxy [96].

Eventually, a campaign of millimetre spectroscopic observations, scanning a wide range of frequencies for an emission line from the carbon dioxide molecule, made a detection [171] and allowed a redshift to be measured, as well as a much more accurate position. The redshift of 5.2, meaning that this object is seen just a billion years after the Big Bang, was surprisingly high. To be as bright as it was seen by SCUBA at such a high redshift, HDF850.1 must be forming stars at a rate of almost 1000 solar masses of new stars each year - nearly a thousand times faster than the rate of star formation in the Milky Way or in a typical optical/near-infrared detected galaxy from the Hubble Deep Fields. Just as surprising was the fact that no optical or near-infrared emission could be seen at the position of HDF850.1, even in the deepest Hubble Space Telescope data available. This galaxy must have so much gas and dust in its interstellar medium that most of its starlight is absorbed by dust and re-emitted in the far-infrared.

The fact that it took fourteen years to get a redshift measurement of HDF850.1 is a salutary warning for all far-infrared and submillimetre astronomers. Sometimes the universe doesn't put your sources where you think they should be, and doesn't make them easily detectable at other wavelengths. HDF850.1 was one of the first five such objects to be found. All the SCUBA surveys combined have found a few hundred more, and other similar instruments that have come along since then have added a few hundred more. Chasing all of those down the way HDF850.1 was eventually dealt with could take a long time indeed. However, catalogs of a few hundred SMGs can now be considered pitifully small. With the Herschel Space Observatory we can now find *hundreds of thousands* of these objects.

7.10   THE HERSCHEL REVOLUTION

The Herschel Space Observatory, shown in Figure 7.10, was launched on 14th May 2009. It was the culmination of many decades of planning - the original ideas for a space based far-infrared telescope date back to the early 1980s, and, once fully approved, it took about ten years to build and get ready for launch. Herschel is one of the largest space astronomy missions ever undertaken by the European Space Agency. Its most obvious unique feature is a huge, 3.5 m diameter, primary mirror, the largest mirror ever launched on an astronomical telescope[3]. Previous infrared astronomy satellites all had much smaller mirrors, with the largest of those being Spitzer at 85 cm. The large mirror allows Herschel to collect nearly seventeen times as much light as Spitzer. When this is combined with more modern detectors, and instruments that extend to longer wavelengths, major advances are possible. You will have seen some of these results in previous chapters but, to my mind at least, it is in the observations of SMGs where Herschel's capabilities are shown most clearly.

Of the three scientific instruments on Herschel, the most important for detecting large numbers of SMGs is SPIRE - the Spectral and Photometric Imaging Receiver [70]. This instrument, developed largely in the UK, uses bolometers to image the sky simultaneously at three separate far-infrared/submm wavelengths - 250, 350 and 500 microns. The combination of Herschel's large mirror, the absence of atmospheric absorption and lack of a large thermal background mean that SPIRE can scan large areas of the sky very quickly. Over its entire career, SPIRE images were taken which covered almost a tenth of the entire sky. The two most important projects in the context of searching for distant SMGs are the HerMES [125] (Herschel Multitier Extragalactic

---

[3]It might be the largest mirror ever launched into orbit, but there are persistent rumours that various spy satellites, looking down rather than up, have still larger mirrors. But nobody is giving details of these, and they're certainly not making them available for astronomers to use.

Figure 7.10 The Herschel Space Observatory.
(Courtesy of ESA http://www.esa.int/ESA.)

Survey) and H-ATLAS (Herschel Astrophysical Terahertz Large Area Survey)[4] surveys [51]. These take two different but complementary approaches to studying distant far-infrared luminous galaxies. The HerMES survey aims to be a little bit like the Hubble Deep Fields, taking sensitive observations in parts of the sky that are already well studied at other wavelengths. It covers a total of about 100 square degrees, with a more recent and less sensitive extension of a further 270 square degrees. The H-ATLAS survey, in contrast, aims for somewhat less sensitive observations but, at 570 square degrees, it covers a much larger area. Both the surveys are in areas of the sky free from contamination by the dust inside our own galaxy, making them ideal for extragalactic astronomy. Between them, HerMES and H-ATLAS have scanned nearly a thousand square degrees, roughly 5% of the sky suitable for extragalactic astronomy, and nearly a thousand times the area covered so far by ground-based surveys in the submillimetre using instruments like SCUBA.

Figure 7.2 gives an impression of just how much of an advance the SPIRE surveys are over what has come before. The SCUBA image of the Hubble Deep Field, shown to scale as an inset, took fifty hours of observing time, and found five galaxies. The much larger HerMES image, a small fraction of the survey as a whole, contains about 700 galaxies and took about eight hours of observations. In total the H-ATLAS and HerMES surveys have detected several hundred thousand SMGs. The days when you could recall the details of every single known SMG are now gone. We now have to look at these objects not as individuals, but as a population, in much the same way that we can look at stars or local galaxies. This really is a revolution, and the final results of this vast new database are still to be seen.

7.11   THE MOST DISTANT DUSTY GALAXIES

One of the things made possible by the huge catalogs of SMGs that are now available thanks to Herschel is that we can start looking for the most exceptional objects. The fact that SPIRE works at three different frequencies simultaneously is a great help. If we look at the typical spectral energy distribution for a star-forming galaxy at low redshift (see Figure 6.5) you will see that such a source will appear 'blue' in the SPIRE bands, meaning that it looks brightest at the shortest wavelengths available. At higher redshifts, though, this colour will change, as shown in Figure 7.11.

A local dusty galaxy will appear brighter in the 250 micron channel than at 350 or 500 microns. At a redshift of three, though, at the upper range of redshifts measured for the SCUBA detected and radio/mid-infrared identified SMGs, it will have roughly the same brightness in the 250 and 350 micron channels. At still higher redshifts, beyond that of HDF850.1, the 500 micron channel will be the brightest. So, as the source moves to higher redshift, the peak of the dust emission, which lies at a wavelength of about 100 microns

---

[4]Any acronym that uses the word Teraherz is clearly just a little bit too contrived.

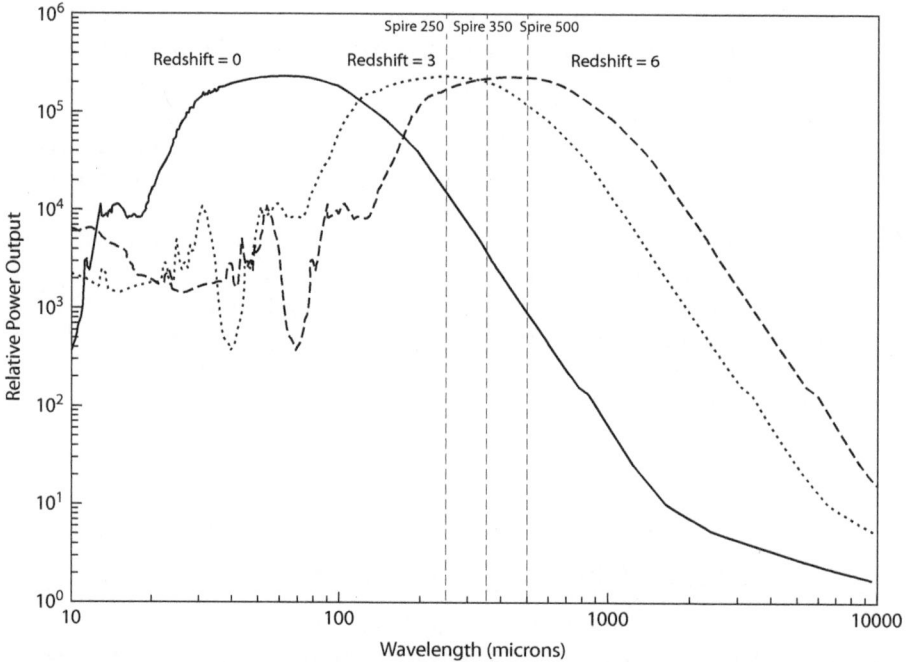

Figure 7.11 How the SED of a dusty galaxy changes with redshift.
This shows how the SED of a dusty star-forming galaxy like Arp220 changes its appearance with redshift. The SPIRE bands are shown with vertical dashed lines, and the galaxy SED is shown at three different redshifts: solid for a redshift of zero, dotted for a redshift of three, and dashed for a redshift of six.

when it is emitted, will be shifted through the SPIRE bands. Any sources that appear red, in other words sources that are brightest at 500 microns, may well lie at redshifts beyond those where SMGs have previously been found.

This colour selection method is actually quite simple to apply to catalogs of SPIRE sources and to SPIRE images. The HerMES team did this to some of the first data that arrived from Herschel and were quite surprised to find more red SPIRE sources than they expected. One of these sources, given the name of HFLS3, was exceptionally red (see in Figure 7.2) and was the subject of a multi-telescope followup campaign to confirm its existence and measure its redshift, which was found to be 6.34 [141], corresponding to a time less than a billion years after the Big Bang. This made it not only the most distant dusty star-forming galaxy currently known, but also put it among the most distant galaxies for which a spectroscopically confirmed redshift is available. Among the many surprising things about HFLS3 is the rate at which it is forming stars - about 2000 solar masses of new stars per year. On its own, HFLS3 is forming stars at a rate comparable to that of all galaxies at a similar redshift that were found in the optical/near-infrared images of the Hubble Deep Field. The mass of dust in this galaxy, about a billion solar masses, is also a problem. The carbon and silicon that makes up this dust has to have been made since the Big Bang, fused from primordial hydrogen and helium in stars. These stars have less than a billion years from the Big Bang until the time at which we see HFLS3 to form, live out their lives, and recycle their enriched material back into the interstellar medium as dust. Clearly stars like our own Sun, which has an expected lifespan of about 10 billion years, are not able to do this in time, so we are beginning to learn something about the very first generation of stars to form in the universe, so called Population 3 stars, even though we have yet to detect any of them directly.

Work is still underway to fully understand the other red sources seen in the HerMES studies which found HFLS3. The search for red Herschel sources is also being extended across the full HerMES and H-ATLAS surveys. A number of other galaxies at redshifts greater than four have already been confirmed, though HFLS3 remains the record holder. Taking these measured redshifts as confirmation that most of these sources are genuinely at redshifts greater than four, we can start to define the properties of this population of dusty, rapidly star-forming galaxies. We can also compare the number of sources we see to the number predicted by the best existing models of galaxy formation. It turns out that none of these models predict the number or brightness of red sources that the Herschel surveys are finding [48]. This presents an interesting problem for those modelling galaxy formation. We can also look at the contribution of these galaxies, which are rare but which are forming stars very rapidly, to the star-formation history of the universe. At first sight, the high redshift Herschel sources only contribute about 10% of the star-formation rate of optical/near-infrared sources at a similar redshift. However, the sources that Herschel can detect in this way have to be very luminous. If you assume that there is an underlying population of similar but fainter objects that Herschel cannot directly detect, you find that these high-redshift dusty star-forming galaxies might contribute as much star formation as the sources found at similar redshifts through optical/near-infrared observations. Once again, it looks as if the optical is only telling us half the story, and our picture of the history of star formation in the universe will only be complete when we also understand what is happening in the far-infrared.

7.12   GALAXIES AS TELESCOPES

While Herschel allows us to detect very distant, very powerful far-infrared and submillimetre galaxies, it is not capable of providing the kind of detailed, high resolution view of these sources that Hubble can provide for the sources it detects. The Hubble Space Telescope has an angular resolution that can be measured in the ten millionths of a degree. Herschel can only resolve structures a hundred times bigger. Also, as we have seen in the previous section, Herschel can only detect the most powerful dusty galaxies at high redshifts, sources like HFLS3 which is forming stars at a massive 3000 solar masses of new stars each year. To be able to understand the processes that drive these sources, to see if their bursts of star formation are triggered by galaxy interactions and mergers, as is the case for their lower redshift, lower luminosity equivalents like Arp220, we need to be able to examine them with much greater precision. New instruments, such as the Atacama Large Millimetre Array (ALMA) are currently being constructed that will be able to do this, but, in the interim, there is another way for us to get a close up view of these sources. This approach also has the potential to tell us something about the mysterious dark matter which makes up much of the mass of the universe.

One of the key predictions of Einstein's General Theory of Relativity is that gravity bends the path of light rays in a process known as gravitational lensing. The degree to which mass bends light is actually quite small, so this effect is only visible when light passes close to a massive object. This can be seen when light passes close to a star like the Sun, which slightly distorts the view of stars close behind it, but it is seen most clearly when light from a distant galaxy passes close to another galaxy or through a cluster of galaxies. An example of the latter can be seen in Figure 7.14, where light from background galaxies passes through the galaxy cluster RCS2 032727-132623, producing a variety of spectacular arcs. These are the distorted images of the lensed background galaxies. Gravitational lensing not only makes the background galaxies brighter, it also spreads their light out over a much larger area. This allows much smaller structures in the lensed background galaxy to be seen than would otherwise be possible. Since the strength of lensing depends on the gravitational mass of material bending the light, you can also use lensing images, like Figure 7.14, to measure the distribution of mass in a galaxy or galaxy cluster. Since most of this mass is made up of dark matter, not the ordinary matter that you, me, the Earth and Sun are made of, determining

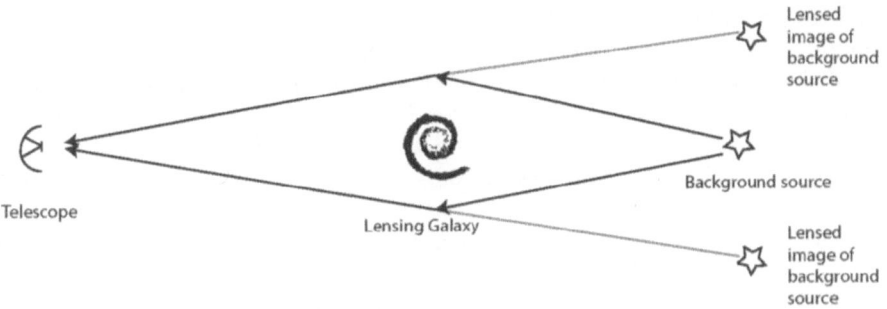

Figure 7.12 **Strong gravitational lensing.**
Diagram showing how galaxy-galaxy lensing produces multiple images of a background source.

the distribution of mass inside objects like galaxies and galaxy clusters is a first step to working out what the dark matter actually is, and how it behaves.

Galaxy clusters can strongly magnify a number of objects behind them, as you can see in Figure 7.14, but they are relatively rare objects. Individual galaxies, which are much more common, can also provide lensing magnification when a distant object lies right behind them. This is known as galaxy-galaxy lensing, and a diagram showing how it functions can be seen in Figure 7.12. A source lying behind a foreground lensing galaxy can, in this way, appear as multiple images. If the geometry is exactly right, the light from the background source is spread out into a ring that surrounds the lensing galaxy. This ring is known as an Einstein ring. As with galaxy cluster lensing, the background source is magnified so it appears brighter than it would otherwise be, and its light is spread out so that smaller scale features can be seen. You can also use the shape of the lensed source to work out the amount and distribution of mass in the lensing galaxy. This means that galaxy-galaxy lensing is potentially a very powerful tool as it can provide a lot of extra information about both the background lensed source, and about the dark matter in the lensing galaxy. If a large number of galaxy-galaxy lenses were available, with the lensing galaxies at a range of redshifts, you could, for example, use them to see how the mass of galaxies grows with time, irrespective of whether that mass is luminous, in the form of stars, or dark matter.

While studies of galaxy-galaxy lenses are potentially very powerful, they are not very easy to find. The efficiency of previous lensing searches, in terms of the number of lenses found compared to the number of candidates studied, has been pretty low. The CLASS lensing survey (Cosmic Lens All Sky Survey), which used radio data, found 22 lensed systems in a survey of 16,000 candidates [116], giving an efficiency a bit better than one in a thousand. Optical searches for strong lenses using the Sloan Digital Sky Survey (SDSS) have found 62 lenses from a sample of about 51,000 candidates [84] for a similar efficiency. Studying the images of potential lensing galaxies in deep Hubble imaging fields has produced twenty lenses from about 9000 candidates, for an efficiency of about one in 500 [56]. The most efficient lens survey to date is probably the Sloan Lens ACS survey (SLACS), which uses a combination of optical spectroscopy from the SDSS and imaging from the Hubble Space Telescope [9]. This has found about 100 lenses with an efficiency of about 68% but relies on both the large existing database of the SDSS, and on expensive HST followup observations. These searches have found a total of about 200 strong galaxy-galaxy lenses [163] which have begun to reveal interesting results about the distribution of mass in the lenses, and about the background lensed sources. What is needed to significantly advance the field is a larger and cleaner sample of galaxy-galaxy gravitational lensing systems, and the central obstacle to this is the low efficiency of most selection methods.

When Herschel was launched, not many people thought it would prove to be useful for gravitational lens searches. However, with hindsight, its potential for pinpointing lenses should have been clear. The key to this is the fact that the background, distant dusty galaxies, discussed in the previous section, are fairly numerous, are quite faint, and have a distinctive colour. Because of this, exceptionally bright sources with the same distinctive red colour are unlikely to be intrinsically bright - sources like this just don't exist. Instead, they appear to be bright because their emission is being magnified by a foreground galaxy.

The first data that arrived from the H-ATLAS survey was searched for candidate gravitational lenses. The H-ATLAS team selected the brightest eleven objects from the 7000 sources detected in the first 15 square degrees of the 570 square degree survey. One of these sources turned out to lie within our own galaxy, one was an unusual radio source (a blazar, see Chapter 6) and four were easily identified as nearby galaxies. That left five remaining candidates. These were all found to have the red colours characteristic of high redshift dusty galaxies. Followup observations of these sources [118] found that every single one of them was the result of strong gravitational lensing, meaning that Herschel has provided us with a near 100% efficient way of finding gravitational lenses. Further observations of these candidates have found some spectacular examples of multiple images and one Einstein Ring - see Figure 7.16.

The exploitation of this technique is still in its early days, but it seems very likely that a large number of lensed systems will be found, stretching out to high redshift for both the lens and the background source. This will be ideal for understanding the assembly of mass over time, and in examining the details of the distant dusty galaxies magnified by gravitational lensing.

7.13 ALMA - THE NEXT REVOLUTION

Herschel isn't the only far-infrared/submm telescope that has proved to be ideal for finding gravitational lenses. The South Pole Telescope (SPT), an instrument designed mainly to study the CMB, uncovered a number of un-

Figure 7.13 Some of the many dishes of ALMA.
(Courtesy of ALMA http://www.almaobservatory.org/.)

usually bright point sources whose colours suggested they might be gravitational lenses. Followup observations with the Atacama Large Millimetre Array (ALMA) showed that these sources were all gravitationally lensed [169]. Even though ALMA was only a quarter completed at the time of the observations, it was still able to produce some of the best images of lensed dusty systems ever seen. Some of these are shown in Figure 7.15.

ALMA will be the next great far-infrared/submillimetre telescope, and the observations of the SPT lenses are just the first glimpse of its capabilities. ALMA will be made up of 66 antennas spread across a plateau 5000 m up in the Atacama desert of Chile (see Figure 7.13). The 66 dishes give it a much greater collecting area than any previous millimetre or submillimetre telescope, and the high, dry location will minimise the effects of atmospheric absorption. ALMA is an interferometer, which means that the light from separate antennas is combined to provide an angular resolution equivalent to that of a single dish that matches the separation of the antennae. Since ALMA's antennae can be up to 16 kilometres apart, this means that finally we will get resolutions in the far-infrared and submillimetre that match, or even exceed, those that optical and near infrared astronomers can achieve with the Hubble Space Telescope.

7.14 CONCLUSIONS

The story of galaxy formation and galaxy evolution is not complete. As we have seen, we now know that our traditional optical and near-infrared observations, with large ground-based telescopes or the Hubble Space Telescope, tell only half of the story, and we need the far-infrared and submillimetre to tell us what happens in the dusty objects responsible for producing the CIB. Only with Herschel has it been possible to find the large numbers of far-infrared sources necessary to really understand what is going on, but our work with the Herschel data is far from complete.

The Herschel space telescope stopped operation at the end of April 2013, after nearly four years in space. It has left a huge legacy of observations, much of which, like the HerMES and H-ATLAS projects, is concerned with the high redshift universe. Meanwhile, new ground-based telescopes such as ALMA, and next-generation instruments such as SCUBA2, the successor to the immensely successful SCUBA, are now coming on stream.

While the story of the dusty, high redshift universe is not yet complete, we now have the tools to fully get to grips with the history of the obscured universe.

Figure 7.14 Gravitational lensing in the galaxy cluster RCS2 032727-132623.
This shows a Hubble Space Telescope optical image of the galaxy cluster RCS2 032727-132623. Galaxies behind this cluster are gravitationally lensed by its large mass, producing the spectacular arc-like structure in the image. (Courtesy of NASA.)

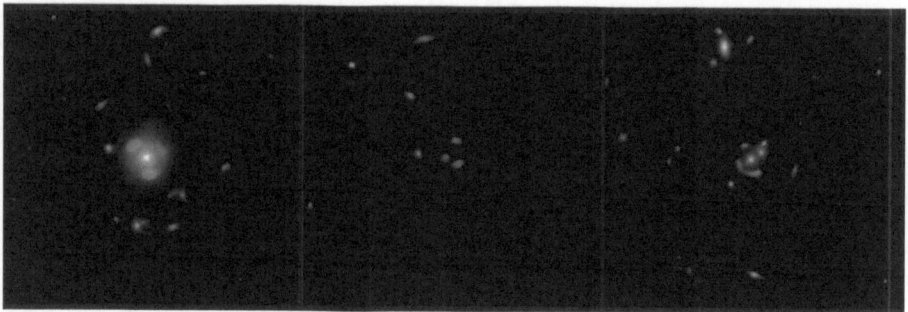

Figure 7.15 ALMA images of the SPT lenses.
The dust emission from the lensed sources is shown here in red, while blue shows HST optical emission from foreground sources and the lenses themselves. As you can see, the lensed objects appear as multiple images, rings and arcs, exactly what you expect to find with galaxy-galaxy gravitational lensing. (Courtesy of ALMA http://www.almaobservatory.org/ [169].)

Figure 7.16 Hubble Space Telescope image of one of the first H-ATLAS lenses.
Light from the lensed source forms the ring around the elliptical galaxy at the centre of the image. (Image used by kind permission of Mattia Negrello and the H-ATLAS Consortium.)

CHAPTER 8

First Heat

8.1   NOT EVEN WRONG

It's 1990, and I'm at my first major academic conference, the American Astronomical Society's big January meeting. There are over two thousand astronomers attending, and there are big talks, major new results, and even a visit from the US Vice President.

My own contribution is a modest poster paper, part of my PhD work. I'm rather proud of it since it presents a way of using observations of what is called the Sunyaev-Zeldovich effect (SZ effect) to answer a puzzle that has emerged over the previous couple of years with the spectrum of the cosmic microwave Background (CMB). The SZ effect is a distortion of the CMB spectrum produced by the hot gas that fills the space between galaxies inside galaxy clusters. A small fraction of the CMB photons scatter off electrons in the hot gas and gain a little energy, slightly distorting the black body spectrum of the CMB in the direction of the cluster.

The problem I think I can solve using the SZ effect is something given the rather grandiose name of the Berkley-Nagoya Excess. This is the result of a sounding rocket experiment launched by a collaboration between Berkley and Nagoya Universities. Sounding rockets aren't capable of reaching orbit, but can ascend above much of the atmosphere and, in the few minutes they are at the peak of their trajectory, they can make astronomical observations impossible from the ground. The Berkley-Nagoya rocket aimed to produce an exact spectrum for the CMB. It flew in 1987 and produced the surprising result that there was excess emission in the submillimetre part of the spectrum, beyond what was expected for a black body at the temperature of the CMB. This has potentially huge cosmological implications, but these will only be clear once the nature of the excess is understood.

There are two widely accepted possibilities. Either the excess is due to dust emission at a temperature a little higher than the CMB, or it is a distortion of the CMB, not unlike the SZ effect but over the whole sky, produced by a large mass of hot gas at an early stage of the universe. My poster shows that there are differences in what you get for the SZ effect depending on which of these possibilities is correct, and shows that this can be measured with the next generation of ground-based instruments.

There is also a third possibility - that the Berkeley-Nagoya Excess doesn't exist, and that the sounding rocket observations were wrong.

One of the big announcements at the meeting is the first set of results from the newly launched COBE satellite (COsmic Background Explorer). Sent into orbit in late 1989, the first results concern the spectrum of the CMB. I walk into the main auditorium of the conference centre, along with about 1000 of my colleagues, to hear the announcement.

After a brief introduction and some technical presentations it is time for the results. The leader of the FIRAS instrument team, the instrument that can measure the CMB spectrum, steps up to the podium, makes a few comments, and then flips on the slide that shows the spectrum of the CMB.

It shows a perfect black body, beautifully fit to the data. There is no sign of the excess seen by the Berkley-Nagoya team, or any spectral distortion many times smaller. The sounding rocket experiment didn't work properly; the data were wrong.

Impressed by the quality of the COBE data, but depressed that my result isn't even wrong, I leave the session, head down to the exhibition hall and remove my poster from the displays.

8.2   INTRODUCTION

In the last chapter we looked at the evolution of galaxies and how observations in the far-infrared and submillimetre are revealing the role of dusty galaxies in the development of the universe as we see it today. The seeds for the formation of these galaxies, though, and for the first generation of stars, came from still earlier stages of the universe. To study them, and to determine the origin and future of the universe itself, we have to look back to still earlier times, and look at the cosmic microwave background (CMB). This is the dull glow that suffuses the entire observable universe with a black body spectrum of photons at a temperature of 2.73 degrees above absolute zero (i.e., 2.73 K).

The discovery of the CMB, together with two other key cosmological observations which we will discuss later, established the Big Bang as our best view of how the universe formed and developed. Starting from those initial results, studies of the CMB have developed until today when, thanks to satellites such as the Wilkinson Microwave Anisotropy Probe (WMAP) and Planck, ground-based CMB experiments such as the South Pole Telescope (SPT) and balloon experiments such as BOOMERanG, we are able to do precision cosmology, with various key parameters of the universe measured to fractions of a percent.

While the CMB was initially discovered, and studied, at radio wavelengths, the latest experiments work mostly in the millimetre and submillimetre regimes. The CMB spectrum itself peaks at a wavelength of about 1.4 mm, just beyond the long wavelength end of what we might call the far-infrared/submillimetre, but it turns out that observations at shorter wavelengths are essential to properly understanding what is going on.

8.3   THE DISCOVERY OF THE CMB

In the 1950s and early 1960s our view of how the universe worked, its history and future, were very different. The dominant theory of the universe was the Steady State Theory, which held that the universe was infinite in both time and space, that it was perpetually expanding, to account for Hubble's Law concerning the motion of galaxies away from each other, and that matter, in the form of hydrogen, was being perpetually created from nothing in the ever-growing empty spaces between galaxies. There were a few problems with this model, principally the relative abundances of hydrogen and helium, which weren't easily accounted for by the theory, but these were not seen as insurmountable. The chief alternative theory was that the universe began with an explosion of space-time at some point in the distant past. This could naturally explain the Hubble Law and the abundances of hydrogen and helium, but didn't fit with the other assumptions of the time. Fred Hoyle, one of the main proponents of the Steady State Theory, derisively called this alternate model the Big Bang during an interview on BBC radio in 1949, a name it has been stuck with ever since.

Arguments about which model was right or wrong raged for many years. Fred Hoyle's Institute of Theoretical Astronomy in Cambridge, for example, faced up against the radio astronomers of the Cambridge University physics department in the Cavendish Laboratory across Madingley Road[1], and there were similar disputes in the rest of the world, wherever cosmologists worked.

All this was to change in 1965 with the discovery of something that only the Big Bang Theory could predict.

In 1964, Arno Penzias and Robert Wilson, both radio astronomers, were working at Bell Labs in New Jersey. They had built one of the most sensitive radio receivers then available, and were using it with a sensitive, six metre horn antenna to study radio signals bounced off the Echo Balloon satellites. These satellites were essentially large, spherical metalized balloons, sent into orbit to act as reflectors for radio signals. The idea was that bouncing signals off these satellites would enable long distance communication from one side of the globe to the other. The signals bounced off the Echo satellites would be quite faint, so to get the system to work it was necessary to build very sensitive instruments and for all interfering sources of noise to be well understood and, if at all possible, eliminated.

Much to their annoyance, despite building a very sensitive receiver system, cooled by liquid helium, and using a telescope designed to eliminate many of the usual sources of interference (Figure 8.1), Penzias and Wilson found

---

[1]The echoes of this dispute remained for many years, long after observations showed that the Big Bang Theory was correct, leading to the Madingley Road being described, at one point, as the widest road in astronomy.

Figure 8.1 The horn antenna used by Penzias and Wilson to discover the CMB.
(Courtesy of NASA.)

persistent residual noise in their system that was a hundred times larger than it should have been. This noise was spread evenly over the whole sky, did not vary from day to night and could not be traced to any source on the Earth, Sun or in our own Galaxy. Despite heroic efforts, one of which was to climb inside the antenna and remove the 'white dielectric substance' left there by nesting pigeons, this uniform source of noise could not be eliminated.

Meanwhile, in one of those odd flukes of scientific history, a different group of radio astronomers, from Princeton University, was conducting a similar experiment less than a hundred miles away. This group, led by Robert Dicke, Jim Peebles and David Wilkinson, had built sensitive instrumentation to search for the residual heat that would have been left behind by the Big Bang if that was the correct theory of the universe. Estimates for the temperature of this radiation had ranged from 50 K to 5 K over the previous twenty years, but there had as yet been no successful attempt to detect it.

The Princeton experiment had yet to go into full operation when a friend of Penzias mentioned a paper by Jim Peebles suggesting that the residual radiation of the Big Bang might be detectable. Such radiation would be highly uniform over the entire sky, and match many of the other characteristics of the persistent noise that Penzias and Wilson had detected. Suddenly they realised that they had discovered something very important indeed. This was the cosmic microwave background (CMB). It was the key piece of evidence that would topple the Steady State Theory, and lead to the general acceptance that the universe began in a Big Bang. It would also lead to the award of a Nobel Prize to Penzias and Wilson in 1978.

8.4   THE EVIDENCE FOR THE BIG BANG

With the discovery of the CMB, the final piece of evidence was in place to show that the Big Bang theory was a far better explanation of what we see in the universe than the older Steady State theory. The three key pieces of evidence were: the Hubble expansion, the primordial ratio of hydrogen to helium and other light elements, and the CMB itself.

The first of these three pieces of evidence to arise was the discovery by Edwin Hubble that galaxies are moving away from each other in such a way that the velocity at which they are moving is directly related to the distance they are apart - double the distance and you double the velocity. The way this works is shown diagrammatically in Figure 8.2, and is often referred to as the Hubble flow, with the constant that relates distance to recession velocity called the Hubble constant.

In the context of Einstein's general relativity theory, the Hubble flow is a result of the space between galaxies expanding. Galaxies can still have separate motions of their own, distinct from the Hubble flow - the Milky Way's motion towards the Andromeda galaxy M31 is an example of such a peculiar velocity - but, as you look on ever larger scales, the Hubble flow becomes the dominant factor. This also provides astronomers with a very useful shortcut to measuring

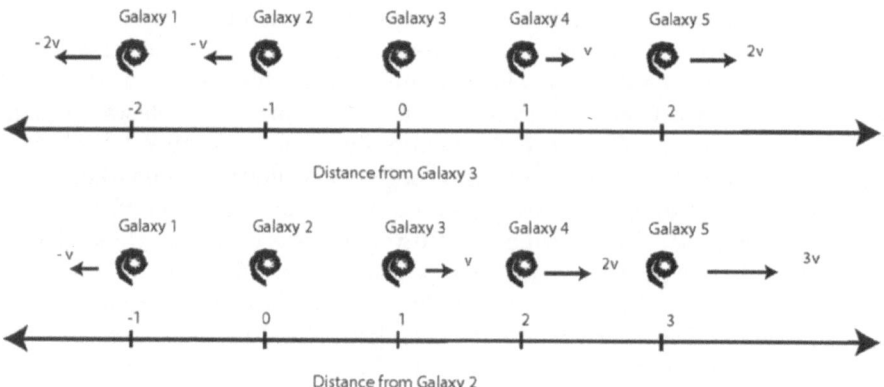

Figure 8.2 The Hubble flow giving the recession velocity of galaxies.
A one-dimensional version of the Hubble flow, showing how recession velocity
can depend directly on distance from the point of view of each galaxy in the
sky. In this case we show the recession velocities as seen from both Galaxy 3
and Galaxy 2, and in each case double, or triple, the distance leads to double,
or triple, the velocity.

the distance of distant galaxies, and is why, especially in the previous chapter,
redshift is often used as if it were synonymous with distance.

The Hubble flow on its own was not evidence for the Big Bang. It might
be tempting to run the flow backwards to a point at which the universe was
much smaller, and all the matter that we currently see in galaxies and in
intergalactic space was crammed into a much smaller, denser volume, but
that was not a unique solution. Just as valid was the idea that the universe
was infinite in both space, size and age, and that the expansion of space
producing the Hubble flow was something that had been going on forever,
within this infinite volume, with matter being continuously created to fill the
ever-expanding void. This idea was proposed by Fred Hoyle and collaborators
in 1948 and, for a time, it became the dominant theory of the universe.

The next set of evidence on the road to the Big Bang is the primordial
abundance of elements and, in particular, the ratio of hydrogen, the simplest
element consisting of just one proton orbited by an electron, to helium, the
next most simple element. Helium comes in two different forms, or isotopes.
The most common has a nucleus made out of two protons and two neutrons,
known as ^{4}He, while a rarer alternate form, known as ^{3}He, has two protons
and one neutron. Both of these are orbited by two electrons. The ratios of
hydrogen to other light elements, such as lithium and beryllium, are also part
of these observations. Primordial abundance ratios are measured by looking
at objects where there has been little stellar activity, since we want to avoid
the effects of hydrogen being fused into helium and other elements by stars.
This is done by looking at the oldest, lowest mass stars in our own galaxy,

which will have undergone relatively little fusion enrichment, or by looking at the interstellar medium of dwarf galaxies, where there has been only a small amount of star formation over the history of the universe.

The results of these observations were that about 25% of the matter in the universe, by mass, is in the form of ^{4}He. Most of the rest is in the form of hydrogen, with trace amounts, of the order of 10^{-10}, of ^{3}He, lithium and beryllium. These ratios presented a problem for the Steady State Model, since in this the production of elements heavier than hydrogen, the process of nucleosynthesis, would only occur in stars, where heavier elements like carbon and oxygen are readily produced. The primordial abundance of these elements was found to be essentially zero, not what the Steady State Model would predict.

The Big Bang, though, provides a natural explanation for the primordial abundances. If you assume the existence of a brief hot and dense phase for the universe, as would occur in the Big Bang when the universe was between 0.1 and 1000 seconds old, you get an environment where fusion can take place which transforms hydrogen into helium and the light elements lithium and beryllium. Heavier elements, like carbon and oxygen, though, will not be formed. Fusion reactions that produce these elements cannot occur through the simple two body collisions that drive nucleosynthesis in the early universe since there are no stable nuclei that contain a total of five or eight protons and neutrons. The production of carbon and oxygen in stars requires three body collisions, which are sufficiently rare, even in the hot dense interior of a star or the early universe, that it takes about ten thousand years for these elements to be produced. The young universe was expanding so quickly that the epoch of nucleosynthesis lasted only a few minutes, so no carbon or oxygen were produced.

The original theory of Big Bang Nucleosynthesis was described by Ralph Alpher, Hans Bethe[2] and Alpher's PhD supervisor, George Gamow in a paper in 1948. This is now often referred to as the $\alpha\beta\gamma$ paper.

Key parameters for Big Bang nucleosynthesis are the relative numbers of neutrons and protons, and the ratio of the number of these particles, together known as baryons, to the number of photons in the early universe. These photons would not just disappear. As the universe expanded and cooled, it would eventually become transparent, and these photons, present in very larger numbers since there had to be about a billion photons for each baryon, would stream freely throughout space. Everything in the universe from that point would, in effect, be sitting in a sea of photons which would form a black body spectrum that gradually cooled with the expansion of the universe.

But where was this background of black body radiation? Its temperature was critically dependent on the age of the universe and on the details of big bang nucleosynthesis. Estimates of the expected temperature of this background radiation ranged from about 5 to 50 K from the 1940s to the 1960s.

---

[2]Hans Bethe was not directly involved in the work that produced this paper, but George Gamow couldn't resist the temptation to produce a paper with an 'alphabetical' author list and so added his friend's name *in absentia*.

Then, in 1965, Penzias and Wilson discovered the CMB, and found it to have a temperature of 2.73 K. With this discovery, the final jigsaw piece was in place, and the Big Bang theory was seen to be the correct description of the early stages of the universe.

8.5   A BRIEF HISTORY OF THE UNIVERSE

The combination of the Hubble Law, Big Bang nucleosynthesis and the CMB allows us to describe the history of the universe from the present time, back to very early stages, if not the actual beginning. The key trick is to run the Hubble Flow backwards, until the universe is very small, very dense, and very hot.

If we run the clock back far enough in this way, we reach temperatures and densities where the behaviour of what we think of as atoms, particles and radiation becomes rather different. As temperature and density increase, the first change we see is that atoms lose their electrons. The resulting state is what is known as a plasma, where electrons and atomic nuclei are mixed together, but the electrons are not bound to any single nucleus. A more familiar example of matter in a plasma state is a flame, where temperatures are high enough to strip some electrons from their parent atoms. The transition of the universe to a plasma state occurs when it was about 300,000 years old. Push the clock back further, and there is a similar transition as the constituents of atomic nuclei, protons and neutrons, are also stripped apart from each other. This occurs about three hundred seconds after the Big Bang. Go back still further, and the protons and neutrons that make up the nuclei of the atoms we see today are themselves broken apart into their fundamental constituents, quarks. These particles are thought to be fundamental and indivisible. The breakdown of protons and neutrons happens at very high energies and densities, corresponding to a temperature of about 10^{29} K, and a time just 10^{-36} seconds after the Big Bang.

This doesn't get us back to the very beginning of the universe, but it is the limit at which we have good reason to believe our physical theories work fairly well. Pushing to much earlier stages in the evolution of the universe would mean we have to confront the key problem that has haunted theoretical physics for much of the last century - that our two hugely successful theories for how the universe works, quantum mechanics and general relativity, need to be unified to describe matter and space at these high temperatures and densities, and we do not yet have a clear idea how that is to be achieved. The early universe is one of the very few places where the implications of such a Theory of Everything might be tested, and we will touch on these matters later in this chapter. For now though, we have reached the point where we can start the clock running forward, and see how the history of our universe took shape.

We start the clock with the universe very hot, very dense, and with the contents of what will end up as atoms today split up into their fundamental constituents - quarks. At this stage the universe was filled with a material that has been called a quark-gluon plasma - gluons are the particles that mediate interactions between quarks, and act as the glue that holds particles like protons and neutrons together - hence the name. All of these particles are in thermal equilibrium, interacting with each other and with everything else that is around, such as photons. This produces a seething mass of fundamental particle physics interactions. In such a thermal equilibrium, particles and antiparticles should be present in equal numbers. This presents us with a problem, since we know that there is more matter than antimatter in the universe today. As the universe expanded and cooled, and moved on from this quark-gluon plasma, something must have happened to create a slight asymmetry between matter and antimatter, so that there is slightly more matter than antimatter. All of the antimatter then annihilated through collisions with particles of matter leaving a little bit of matter left behind. When matter and antimatter annihilate they produce photons, so the level of asymmetry between matter and antimatter can be found by comparing the number of photons in the universe, found in the CMB, to the number of baryons, - i.e., protons and neutrons - that survived this mass annihilation. There are about a billion photons for each baryon, so the matter-antimatter asymmetry is very small, but our very existence depends on it.

The physics behind the matter-antimatter asymmetry is not fully clear, and is tied up with what are called grand unified theories of particle physics that attempt to unify the strong nuclear force that is carried by gluons with the other force of particle physics, the electroweak force. The idea is that the particles that mediate interactions between baryons and leptons - particles like electrons and neutrinos that interact via the weak nuclear force - decay in such a way that they produce slightly more matter than antimatter [126]. The process by which this happens as the universe expands and cools is called baryogenesis, since this produces the building blocks which make all the baryons we see in the universe today. The baryogenesis phase of the universe is thought to have ended when its temperature dropped below about 10^{27} K, just 10^{-36} seconds after the Big Bang.

As the universe continues to expand and cool, its temperature passes a number of other critical values in the context of particle physics. At a temperature of about 10^{15} K, and about 10^{-12} seconds after the Big Bang [24], the weak nuclear force and the electromagnetic force become distinct, as it becomes possible to tell the difference between massless photons and the massive W and Z particles that mediate the weak force. Then, at a temperature of about 10^{12} K, and a time of about 10^{-5} seconds (i.e., 10 microseconds) after the Big Bang, temperatures are low enough that quarks can no longer move freely in the quark-gluon plasma. Instead, they become confined, and freeze out into protons and neutrons, the particles that make up the nuclei of the atoms we see today. The cosmological implications of these phase transitions, beyond producing the particles we are familiar with today, is unclear, so we won't dwell on these possibilities at the moment, but will come back to one, inflation, later on.

From the ages of 10 microseconds to 100 seconds, things didn't change very much, but by the time the universe was a little over a minute old, the universe was so cold that protons and neutrons could start binding together to form atomic nuclei more complex than the single proton we find at the centre of a hydrogen atom. This is the epoch of nucleosynthesis. The observed abundances of the different light elements that emerged from this, as we have seen, provided the earliest indication that the universe began in a hot, dense Big Bang. Nucleosynthesis was over when the universe reached an age of about 300 seconds.

The next major transition in the universe came when it had cooled so much that electrons could no longer be prevented from combining with the nuclei produced during nucleosynthesis, and they combined into the first atoms. This corresponds to a temperature of about 6000 K, and took place about 300,000 years after the Big Bang. This stage, known as the epoch of recombination[3], saw the universe change from being filled with plasma, which is opaque to photons, into a gas of neutral atoms, making it transparent. The photons we see today in the CMB were released at this point, and could stream freely across the universe for the rest of time. What we see when we observe the CMB today is in fact an image of the universe when it was just 300,000 years old.

A common feature in this very brief history of the universe, summarised in Figure 8.3, is the concept of phase transitions, where the expanding and cooling universe drops to a temperature low enough for some process to stop operating, such as the ionisation of atoms that kept the universe filled with plasma until the epoch of recombination. Phase transitions are not exotic phenomena, confined to high energy physics labs or the early universe; instead they are all around us. An example is water cooling so much that its individual molecules are no longer moving fast enough to overcome the forces that would otherwise bind them together in a stationary, crystalline structure. When this happens, water freezes into ice. Many of the same rules that govern phase transitions between ice, water and steam also govern the physics of the early universe. However, the detailed effects of phase transitions in the early universe depend on yet-to-be fully-understood laws of physics. By observing the end results of the early evolution of the universe, through observations of the CMB, for example, we can get a better idea of what those fundamental laws of physics might be.

---

[3]Technically, since electrons and atomic nuclei had never before combined together to make atoms, this should be known as the epoch of combination, not recombination, but for some reason cosmology has got stuck with the term 'recombination' for this process.

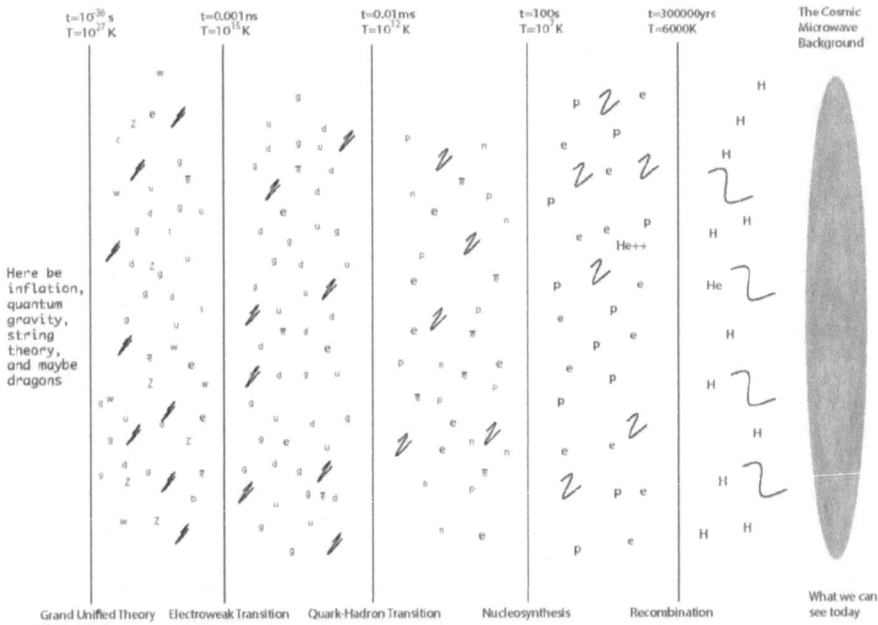

Figure 8.3 Phases in the early history of the universe.

The times and temperatures of the important phases in the early universe are shown here, along with a representation of the constituents, from the hydrogen and helium atoms we see today, through the ionised hydrogen (protons) and helium atoms before recombination, the separate protons, neutrons, electrons and positrons (the antimatter equivalent of electrons, indicated as ē) before nucleosynthesis, the up and down quarks and gluons before the quark-hadron transition, to the w and Z particles and more exotic quarks that existed after baryogenesis and the grand unified theory transition. This all takes place amid a sea of photons, indicated by the sqiggles, that get ever more energetic the further we look back in time. Other particles, like neutrinos and heavier relatives of electrons known as muons and tauons, are not shown for clarity. (CMB map at the right courtesy of ESA.)

8.6 THE SEARCH FOR ANISOTROPY

Once the CMB was detected as a uniform glow across the entire sky, the race was on to look for variations in its temperature from place to place. A perfectly smooth, uniform CMB could be described as isotropic - it looks the same in all directions. Variations from perfect smoothness are thus known as anisotropies. Such CMB anisotropies could be produced in a number of ways. They might arise when the CMB was produced at the epoch of recombination, or they might result from processes that occurred since the photons of the CMB were released to stream freely through the universe.

The first anisotropy discovered actually has a very local cause, and arises because the Earth and Solar System are moving relative to the photons of the CMB. This produces a simple Doppler effect, whereby the photons in the direction towards which we are travelling appear to have a little more energy, and thus appear to have a slightly higher temperature, and vice versa for those photons in the direction from which we are travelling. This produces a characteristic distribution of temperature difference across the sky known as a dipole, which was detected in a series of experiments using balloons-borne instruments [38], [76] or observations from a high altitude U2 plane [156] during the 1970s. The Solar System turns out to be traveling at a velocity of about 270 km/s relative to the CMB, which produces a temperature increase of about 0.0034 K (3.4 mK) in the direction towards which we are moving.

This dipole, which is shown in the middle part of Figure 8.4, is an example of a secondary anisotropy, which have nothing to do with the processes that produced the CMB in the early universe. The search for primary anisotropies took much longer to achieve success, but the scientific return from their discovery would be much greater. The reason for this is two-fold. Firstly, primary anisotropies can be used to measure the physical properties of the universe when it was just 300,000 years old and, by inference from these properties, to determine some of the physics that drove the earliest stages of the universe. Secondly, fluctuations in the temperature of the CMB also correspond to variations in the density of matter from place to place at the epoch of recombination. Regions where the matter density was higher would, through the eons of cosmic time that followed recombination, grow through gravitational attraction to form the stars, galaxies and galaxy clusters we see in the universe today. CMB anisotropies are thus a link between today's universe, and the fundamental physics that sowed the seeds that made it the way it is.

8.7 NOISE IN THE DARKNESS

The fractional variation in temperature across the sky in the CMB dipole is a little bit larger than one part in a thousand. The variations that come from the epoch of recombination itself, the primary anisotropies, are much smaller than that, less than one part in a hundred thousand. If we think of this as the roughness of the surface of a sphere the size of the Earth, with a radius of

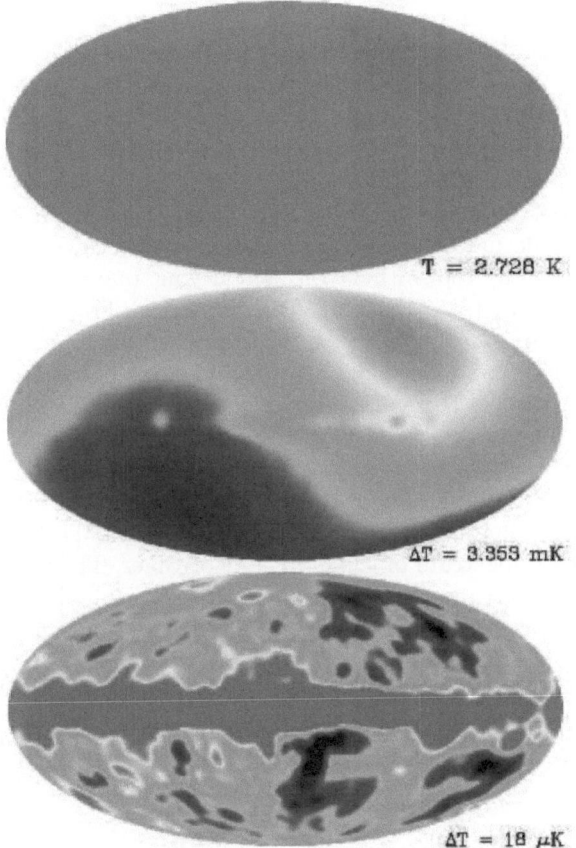

T = 2.728 K

ΔT = 3.353 mK

ΔT = 18 μK

Figure 8.4 The CMB as seen by the COBE satellite.

The top image shows the uniform, 2.728 K temperature of the CMB over the entire sky. As you can see, there are no significant deviations in temperature to be seen. The second image shows the small (about one part in a thousand) deviation in temperature due to our motion relative to the CMB, the CMB dipole. This can be seen as an increase in temperature in the direction we are travelling (upper right in these figures) and a decrease in the direction away from travel (lower left). The bottom figure shows the deviations in CMB temperature that are found once this dipole distortion is removed, finally revealing the primary anisotropies that are the imprints of density variations in the universe at the epoch of recombination. These variations amount to less than one part in 10^5, in terms of both temperature and density, but are the seeds from which all structure in the universe grew. These images show the entire sky, using a projection similar to what you would get if you cut a tennis ball in half and then squashed the two hemispheres flat. They are also aligned so that the plane of our own Galaxy, which is the brightest contaminating foreground source for CMB studies, lies across the centre of each image. The odd lumps you can see in the middle image, and the bright band you can see across the bottom image, are thus nothing to do with the CMB, but are in fact a result of local emission from the Galaxy. (Courtesy of NASA.)

6400 km, then the primary anisotropies would correspond to surface features less than 100 m high. The first thing to note about the primary anisotropies, then, is that they are very small.

The second thing to note is that they are random. There are no clear coherent structures in the CMB, like stars, galaxies or planets. Instead, there are just random fluctuations about the mean temperature value. However, these random fluctuations are not the same on all scales. They are different from the pure 'white' noise that you hear if you tune an FM or AM radio to a blank channel. Instead, there are certain size scales on the sky where the random fluctuations are stronger, and certain scales where they are weaker. If we continue with the sound analogy, we could think of looking at the strength of CMB fluctuations on small scales as corresponding to high notes, or high frequencies since you can have a lot of small fluctuations around the entire sky. Conversely, their strength on large scales corresponds to low notes, or low frequencies. Observations of the CMB anisotropies aim to determine the strength of these fluctuations on different size scales. This is known as measuring the power spectrum of the CMB anisotropies. The results are a bit like looking at the spectrum analyser you can sometimes find on a hi-fi system or music playing app, but in this case we're looking at the entire sky, not a single moment in a piece of music.

The strength of CMB fluctuations on different scales is determined by factors related to the properties of the plasma that filled the universe at the time of recombination, on the expansion rate of the universe, and on the physics of the earlier stages of the universe that provided the seeds for these fluctuations.

What we see today as regions of higher or lower temperature in the CMB are the result of enhancements, or reductions, in the density of hydrogen and helium nuclei, dark matter particles, and the number of photons at a particular place on the sky at the epoch of recombination. The matter density is dominated by dark matter particles. These only interact with other particles through gravity. The hydrogen and helium nuclei, and especially the electrons associated with them, interact strongly with the photons. The photons themselves try to stream freely through the plasma but they are continually bumping into electrons. When these collisions happen, the photons push the particles in the plasma apart. There are thus two competing effects at work at the epoch of recombination: firstly there is gravity, which tends to pull matter into regions of higher density. Secondly there is pressure, coming from the interactions between photons and electrons. While gravity pulls matter together, this pressure acts to push matter apart.

A similar, but more familiar, arrangement like this is when a weight is suspended by a spring. Gravity pulls the weight down, but tension in the spring opposes this. The weight ends up bouncing up and down. In just the same way, the material inside the universe, prior to the epoch of recombination, vibrates, filling all of space with sound waves. However, unlike the weight on a spring, whose oscillations will carry on forever if we ignore inevitable energy

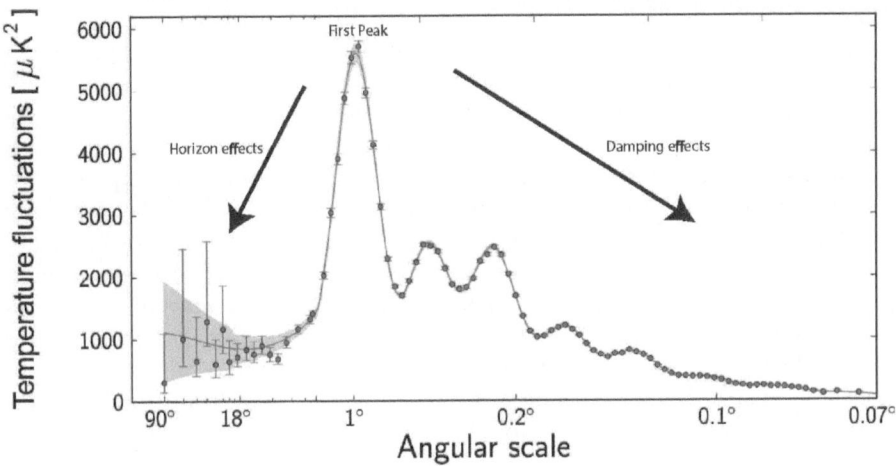

Figure 8.5 The power spectrum of the CMB.
This plot shows the power spectrum of the CMB from theoretical models (line) and as observed by the Planck satellite (points). As can be seen, the strength of the fluctuations falls off towards small scales, thanks to damping effects, and on large scales, thanks to the horizon size, leaving a prominent peak on scales of about a degree, as well as smaller peaks at smaller scales. By matching the observed data to models such as the line shown here, we can determine a variety of key features of the universe we inhabit. Another aspect of this plot is the remarkable degree to which the theoretical model and the latest Planck data agree with each other [132]. (Courtesy of ESA: http://www.esa.int/ESA.)

losses, there is a time limit to how long the plasma oscillations in the young universe can last. This is because the universe is expanding and cooling, so the plasma will eventually recombine to make neutral atoms. At this point matter and radiation decouple, and the photons can at last fly freely across the universe to make up the CMB, carrying with them the imprint of the hot and cold regions produced by these acoustic plasma oscillations.

On small scales, there is a tendency for photons to leak out of the oscillating regions, and to diffuse away. High frequency oscillations, corresponding to small scales, are suppressed by this, in a process called Silk Damping, named after the astrophysicist Joe Silk who first came up with the mechanism [153].

Oscillations on the largest scales are suppressed by a different effect, which results from the finite age of the universe. Since the speed of light, at 300,000 km/s, is the ultimate speed limit in the universe, regions separated by a distance greater than this speed times the age of the universe can have no effect on each other. They are, to use the technical term, beyond the horizon from each other. The tendency for the density variations we see in the CMB to grow under the effect of gravity cannot act over distances larger than the horizon at recombination. CMB fluctuations larger than this size are suppressed.

Given these two effects, we might expect the power spectrum of CMB anisotropies to reach a peak at the scale of the horizon size at recombination, and that is in fact what we find. The size of the horizon at recombination depends on a number of things, but most importantly on the expansion rate of the universe, so this sets the frequency at which we find the peak of the CMB power spectrum. The strength of the peak depends on other factors, especially the coupling of normal matter, in the form of hydrogen and helium nuclei, to the photons. The stronger the peak, the more normal matter there is in the universe. Measuring the frequency and strength of other peaks in the CMB fluctuation spectrum provides similar insights into other key parameters of the universe. Figure 8.5 shows both a theoretical model of the CMB power spectrum, and also the results of the latest CMB observations by the Planck satellite [132].

This theory of CMB anisotropies, and the realisation of how important they were to understanding how the universe works, was developed in the decades that followed Penzias and Wilson's discovery. But it was not until the early 1990s, more than 25 years after the CMB was discovered, that primary CMB anisotropies were detected.

8.8   THE NUMBERS THAT MAKE A UNIVERSE

What are the key numbers that determine the nature of the universe we inhabit? It turns out that there actually aren't very many of them, and most can be determined from the CMB or from the combination of CMB data with other, complementary observations. There are also a number of other, more technical, parameters that emerge from this analysis, but the key numbers, that make the universe what we see today, include the following:

- The Hubble constant - given the symbol H_0

This gives us the current rate of expansion of the universe. Together with other parameters, H_0 can be used to find the age and size of the universe.

- The curvature of the universe - given the symbol Ω

In the context of general relativity we can think of space, and the universe as a whole, as being curved. The universe can have a number of different curvatures. It might be curved round on itself, like the inside of a ball, and would mean that the universe will eventually recollapse on itself, maybe triggering a new Big Bang. This corresponds to space having positive curvature, with Ω having a value greater than 1. Alternatively, the universe might expand forever, leading to the opposite effect, negative curvature, with Ω having a value less than 1. Balanced between these two alternatives is the case where the universe is flat, with no curvature at all and Ω equal to 1. The value of Ω is determined by the density of the universe since it is gravity that acts to curve space. Above a certain critical density, there is enough gravity to, at some future time, stop the expansion, and reverse it into a collapse leading to a big crunch. Working out what value Ω actually has will tell us the future of the universe.

• The contribution of normal matter to the universe - given the symbol Ω_b

Several different things contribute to the density of the universe and thus the overall value of Ω. Major contributors include normal matter, the kind of stuff we're made of, which includes protons and neutrons. The part of Ω contributed by normal matter is known as Ω_b, where the b stands for baryons, the term for particles like protons and neutrons. Knowing Ω_b will tell us how important, or unimportant on a cosmic scale, is the material that we are made of.

• The contribution of dark matter to the universe - given the symbol Ω_d

The next contributor to Ω is the dark matter that interacts with normal, baryonic matter, through gravity. We can see its gravitational effects on galaxies and other large structures in the universe, but we don't yet know what the dark matter is made of. It is thought to be made up of a class of subatomic particle that reacts only weakly with normal matter. Determining the amount of dark matter in the universe through a measurement of Ω_d will help sort out what it is made of, and also allow us to better understand the role it plays in galaxy formation and evolution.

• The contribution of dark energy to the universe - given the symbol Ω_Λ

One of the surprises of recent decades in cosmology has been the observation that there is something else, in addition to normal and dark matter, that contributes to the density of the universe. This is something that has been given the name dark energy, but that name is really just a label for something that is not well understood. Its effect is as if empty space could exert a pressure on not only things within it, but on space itself. Such a material would cause the expansion rate of the universe to increase over time. This acceleration in the expansion rate was discovered when astronomers started monitoring the behaviour of distant supernovae, and resulted in the award of the 2011 Nobel Prize to Saul Perlmutter, Brian Schmidt and Adam Reiss. We still don't know the origin of this dark energy, or how it behaves over time (see Chapter 9 for details of the Euclid mission, which should lead to improvements in our understanding of dark energy), but CMB observations can determine its contribution to the overall density of the universe. This contribution is called Ω_Λ.

The overall value of Ω is the sum of all the types of material that contribute to it, so Ω is the sum of Ω_b, Ω_d and Ω_Λ.

• The Spectrum of initial perturbations - given the symbol n

While the other parameters on this list are all to do with the expansion of the universe and what drives its behaviour, the final one deals with what happened much earlier in the Big Bang. The small fluctuations we see in the CMB did not arise spontaneously at the epoch of recombination. Instead, they must have arisen from fluctuations seeded into the universe at still earlier epochs. The natural way for this to happen is through the inherent randomness of quantum mechanics. This will give rise to what we could think of as noise, random differences from one place to another in the density of the very early universe. This noise will be white, in the sense that it has no favoured scale. The spectrum of these initial perturbations can be described by a simple number, known as n. If the initial perturbations are genuinely scale free, then the value of n will be precisely 1. Any deviation from that value will be a signature of physical processes at the very earliest stages of the universe.

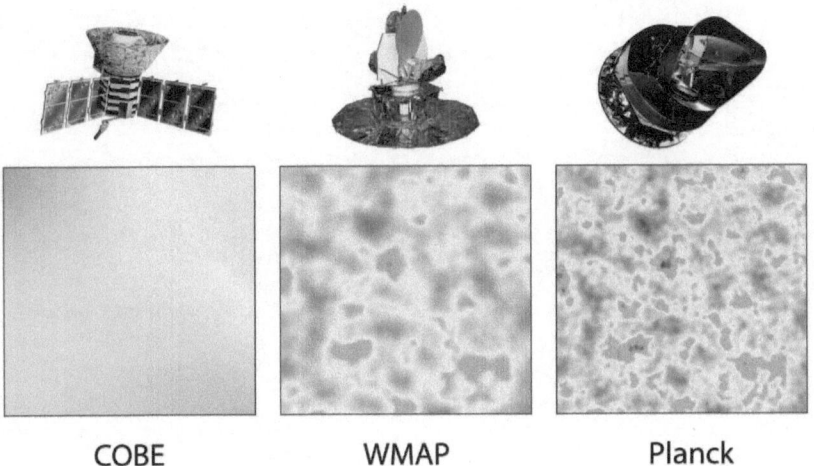

Figure 8.6 Comparison of the resolving power of the CMB satellites COBE, WMAP and Planck.
These satellites worked at progressively higher resolutions, and so can resolve ever smaller structures. (Courtesy of NASA.)

Figure 8.7 The all-sky view of the CMB from the Planck mission. (Courtesy of ESA http://www.esa.int/ESA.)

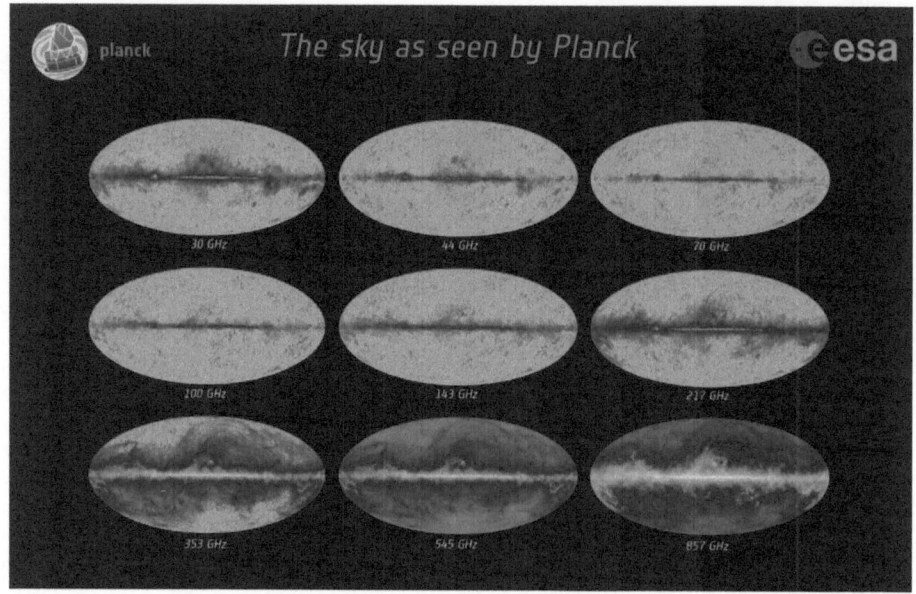

Figure 8.8 All-sky maps from the nine different frequency bands of Planck.
(Courtesy of ESA http://www.esa.int/ESA.)

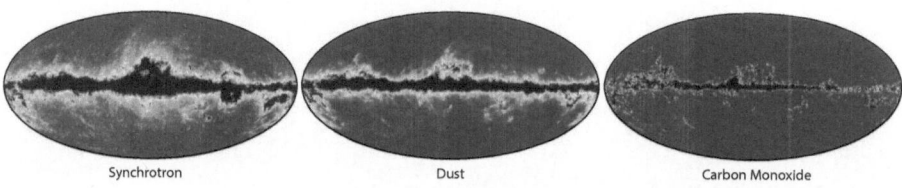

Figure 8.9 All-sky maps of the three main foregrounds contaminating the Planck view of the CMB.
(Courtesy of ESA http://www.esa.int/ESA.)

8.9   CMB OBSERVATIONS

As you might guess from the 25 years between the discovery of the CMB and the first successful detection of primary anisotropies, these observations are not easy. The thermal spectrum of the CMB peaks at a wavelength of 1.4 mm, but much of the observational work on CMB anisotropies has, until very recently, been conducted at rather longer radio wavelengths. The reason for this is that, despite the extra signal that you get by working close to the peak of the CMB, all the observational difficulties discussed in Chapter 2, such as strong backgrounds and poor detector sensitivities, have meant that you could actually do better in the radio. Many observations of the CMB, therefore, have been conducted at radio or millimetre wavelengths. While such observations were, eventually, able to achieve the necessary sensitivity, working in the radio brings the additional problem of resolution, since, at these wavelengths, you would need an impossibly large antenna to be able to measure structures on small angular scales. You can see this effect in Figure 8.6, which compares observations of the same piece of sky by the three generations of CMB satellites that have so far been sent into space.

There is also another factor at work here. The CMB is, almost by definition, the oldest light in the universe, and thus it lies further away from all other sources of radiation. To be able to get an accurate measurement of what the CMB is doing, it is necessary to remove the contributions of every other source of electromagnetic emission at the wavelength of your observation. This is known as foreground subtraction since, from the point of view of the CMB, everything else in the universe is in the foreground.

There are a range of different processes producing foreground emission that will cause problems for CMB observations. At long radio wavelengths, synchrotron radiation is the main contaminant. This is radiation produced when free electrons in the interstellar and intergalactic medium spiral around magnetic fields. Our own galaxy is a rich source of synchrotron radiation, and this produces the distinct galactic band in the CMB images from the COBE satellite in Figure 8.4. Other galaxies, and especially AGN, can also be strong synchrotron emitters. At shorter wavelengths, thermal emission from dust is a major problem. You already know from what has been said in Chapters 6 and 7 that our own and other galaxies are strong dust emitters, so this is another foreground that must be removed. Finally, close to the wavelength at which the CMB is brightest, there is a third source of foreground contamination. This comes from an emission line in the carbon-monoxide molecule, CO, which can produce a strong but fairly localised foreground.

These are the three strongest foreground contaminants for CMB observations. Helpfully, when you look at how their strength changes with wavelength, they all behave differently, both from each other and from the CMB. Given enough observations at different wavelengths, then, we should be able to separate out the different foreground components, then subtract them from images of the sky to produce a pure, CMB only, map.

Early CMB observations were conducted from the ground using radio telescopes of various designs. After the early success of the CMB dipole detection in the 1970s, it became clear that space-based observations would be needed for a full understanding of the anisotropies. Not only could observations in space escape the usual problems of the atmosphere and, for radio observations, interference from terrestrial, man-made sources, but going into space would also be the only way to get a complete map of the entire sky with the same instrument. This can't be done from the ground since the Earth gets in the way, cutting out half of the sky.

The COBE satellite (Cosmic Background Explorer, the leftmost spacecraft in Figure 8.6) was the first such mission, and was a roaring success. It provided the first detection of primary anisotropies as well as several other significant results discussed here and in Chapter 8. However, COBE wasn't able to resolve structures small enough to be able to see the first peak in the anisotropy power spectrum.

The years following the COBE results were a race to measure the first peak in the CMB anisotropy spectrum. The next space mission, WMAP (Wilkinson Microwave Anisotropy Probe, the middle spacecraft in Figure 8.6) was proposed in 1995, not long after the final COBE results were announced. It would certainly have the capabilities to detect the first peak, and much more, but it wouldn't be launched until 2001. There was thus a window of opportunity for ground- and balloon-based experiments to make progress before WMAP was operational. A series of experiments managed to do this just before WMAP was launched, including the BOOMERanG [106] and MAXIMA balloons [74], and the MAT/TOCO [121] and Saskatoon [120] ground-based instruments. When WMAP was finally launched, it confirmed and massively improved on these earlier results, and added further insights at smaller scales.

The best, however, was yet to come.

8.10   THE PLANCK REVOLUTION

At the same time that the WMAP mission was being proposed to NASA, two different missions, called COBRAS and SAMBA, were proposed to the European Space Agency (ESA) to make observations of the CMB. One of these, COBRAS, was broadly similar to WMAP, working in the radio, and with limited angular resolution. The second, SAMBA, was much more innovative, and thus risky, working at shorter wavelengths, stretching into the far-infrared and submillimetre, and able to resolve structures six times smaller. These missions had similar science goals, so ESA suggested that the two teams should combine their efforts and propose a joint mission, with sensitivity stretching from the radio to the far-infrared. The eventual mission, which suffered for some years under the hideous name COBRAS/SAMBA, was approved in 1996 and renamed Planck. By covering nine different wavelengths, from the radio to far-infrared, including the peak of the CMB energy distribution at 1.4 mm, Planck could measure and remove all the different contaminating foregrounds.

This gave it the capability of bringing a new era of precision to the field of cosmology, measuring a wide range of cosmological parameters to accuracies better than 1%, including those listed above.

Planck was launched together with the Herschel Space Observatory on 14th May 2009. It then spent the following four years continuously scanning the sky, producing several independent all sky surveys in each of its nine channels. The resulting maps were at higher resolution than any previous all-sky CMB survey (see Figure 8.6) and also reached higher sensitivities. To get the best out of the data, though, the nine different frequencies have to be combined in such a way as to subtract out the contributions from the foreground universe, leaving only the CMB signal[4].

Figure 8.8 shows the nine different all-sky maps currently produced by Planck and based on the first two all-sky surveys, all aligned so that the plane of our own galaxy goes across the centre of each image [130]. As you can see, the structures visible on the sky in the different frequency maps are all subtly different from each other. At a frequency of 30 GHz (a wavelength of 1 cm), the sky is dominated by synchrotron emission in our own galaxy. As we move to higher frequencies, and shorter wavelengths, the synchrotron emission becomes weaker. However, emission from dust in our own galaxy gets stronger at shorter wavelengths, with the 857 GHz image (a wavelength of 350 microns, well into the far-infrared) completely dominated by dust. Emission from the carbon monoxide molecule also contributes to the 100 GHz (3 mm wavelength), 217 GHz (1.4 mm wavelength) and 353 GHz (850 micron wavelength) channels [131]. This is why the 143 GHz map appears to have the least structure in it once you get away from the galactic emission across the centre of the image.

Working out how to combine the nine different maps in such a way as to separate out the three main foreground contaminants, leaving a CMB-only map, was quite a difficult task, taking many person-years of effort by the Planck team. The final results for the foreground components are shown in Figure 8.9, while the resulting all sky map of the CMB fluctuations is shown in Figure 8.7.

Further processing is needed to go from this map to the power spectrum of the anisotropies shown in Figure 8.5, and then some more work is needed to compare the observed values of the power spectrum to those produced by different values of the key cosmological parameters. Once all that is done, the Planck data allows us to calculate the values of these parameters to accuracies better than one percent. This is a huge advance compared the the status of cosmology when I started my PhD. The joke then, not without good reason, used to be that if a cosmologist managed to get something right to within a factor of 2 they were doing well. With the results from Planck, we now have to get things right to better than one part in a hundred. The era of precision cosmology has arrived.

---

[4]There is of course an awful lot of astrophysics in these foreground signals, so many astronomers, myself included, have spent a lot of time looking in the dustbins of the CMB cosmologists to produce useful results.

Of course the Planck data did not arrive in a vacuum. As well as the results from the previous satellite WMAP, new generations of ground-based CMB experiments were under way using more recent detector technologies than Planck. These experiments include the South Pole Telescope (SPT) [148], and the Atacama Cosmology Telescope (ACT) [43]. These ground-based experiments scooped the Planck project for results on smaller scale anisotropies (e.g., [89]), much to the annoyance of some members of the Planck team. They also provided a useful independent cross check on the Planck data. However, since these telescopes are ground-based, they are unable to cover the entire sky, which is necessary for a full understanding of the CMB power spectrum.

8.11   THE PLANCK VIEW OF THE UNIVERSE

What are the precision results that have come from Planck?

For the key numbers discussed above, Planck, when combined with data from other cosmological studies, produces the following results [133]:

- The Hubble constant, $H_0 = 67.3$. Bearing in mind that even in the early 1990s, the value of the Hubble constant was uncertain to a factor of 2, this is a huge advance. Interestingly, precision measurements of the Hubble constant using the Hubble Space Telescope produce a somewhat higher value, closer to 75. Given the uncertainties of these precision measurements, this disagreement is on the edge of being interesting. Whether this disagreement is real is still unclear. Even if it is real, it might be saying more about our lack of understanding of various classes of pulsating stars than suggesting that there is a problem with our understanding of cosmology and the CMB. Further observations, and further analysis of the Planck data, are needed to clarify this question.

- The curvature of the Universe, $\Omega = 1$. This means that the universe is geometrically flat and that it will expand forever. This result confirms results from previous studies and matches a number of theoretical assumptions.

- The normal matter content of the universe, $\Omega_b = 0.049$. This means that normal baryonic matter, the kind of material that we are made of, contributes less than 5% of the density of the universe. Everything we see in the sky, and everything discussed in other chapters of this book, are really just the froth on top of the dark matter that really drives the growth of structure in the universe through its gravitational attraction.

- The dark matter content of the universe, $\Omega_d = 0.268$. This is the flip side of the normal matter content, showing that there is more than five times as much dark matter in the universe as normal matter.

- The dark energy content of the universe, $\Omega_\Lambda = 0.683$. This refines previous results on dark energy, the mysterious substance responsible for the pressure that accelerates the expansion rate of the universe. As expected, dark energy is responsible for most of the density of the universe at the current time, and it is rather embarrassing for cosmologists that we know so little about it. These results from Planck set the scene for future projects, including the future ESA satellite mission called Euclid, that will study dark energy in much greater detail.

• The initial perturbation spectrum, $n = 0.9603$. This value seems very close to 1, the value you would expect for scale free perturbations coming from quantum fluctuations in the early universe. However, the precision with which Planck has determined n means that, while the difference from a value of 1 is very small, the chances of the value actually being 1 given this data are very small indeed. This deviation from strict scale invariance is a signature that the universe went through a period of very rapid expansion at very early times. This process, called inflation, is likely to be driven by the fundamental physics that unifies general relativity and quantum mechanics. Planck's precise determination of n is the first step to comparing real observations of the universe to the many theoretical models for inflation that rely on string theory and other approaches to a theory of everything.

As well as these headline numbers, Planck also provides much more precise numbers for a whole range of other parameters necessary to describe the early universe, including limits on any variation in the fundamental constants of physics, changes in dark matter properties over time, and on primordial magnetic fields. If evidence had been found for any of these it would have been a major surprise.

Some have claimed that the view of the universe that has emerged from the Planck data is of a 'maximally boring universe' that matches up to all our previous data and many of our theoretical expectations. You might take that as an indication that Planck hasn't contributed anything new, but that would be very wrong. What Planck has shown is that all the work done in cosmology since the discovery of the CMB has been right, and that we are genuinely making progress in understanding how the universe works.

8.12   CMB ANOMALIES

There are a couple of odd things that were hinted at in the WMAP maps of the CMB that Planck has confirmed. These are known as the 'cold spot' and the 'axis of evil' and are highlighted in Figure 8.10, which shows the same CMB map as in Figure 8.7, but with the anomalies indicated.

The cold spot is a region on the CMB that is unusually cold - about 70 millionths of a K colder than the temperature of the CMB. The average CMB temperature variation is about 18 millionths of a K. This means that there is less than a one in fifty chance of the cold spot occurring if it were produced by the same processes that led to all the other CMB anisotropies [40]. A number

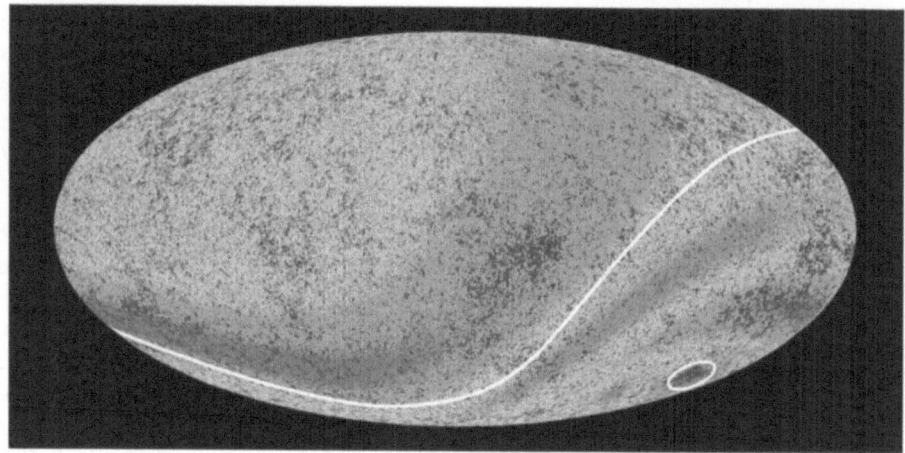

Figure 8.10 The Planck CMB map, highlighting the anomalies.
The cold spot is indicated by the ring in the bottom right. The white line shows the orbital plane of planets in our own solar system, the ecliptic, and you can see the 'axis of evil' in that to the southern side of this line the CMB is generally warmer than the north. (Courtesy of ESA; http://www.esa.int/ESA.)

of possible origins for the cold spot have been suggested, including the effects of a large void in the distribution of galaxies towards the cold spot, and the effects of phase changes in the very early universe.

The 'axis of evil', as it has been called [54], is perhaps even more odd since it is something seen in the CMB that is apparently related to the orientation of the plane of the ecliptic - the orbital plane of our own Solar System. On one side of the ecliptic the CMB sky is, on average, warmer than on the other side. This isn't something that anybody expected. Suggested explanations include large scale magnetic fields in the early universe, the possibility that the universe as a whole is rotating, or the possibility that this is a signature of some new physical process that we don't understand.

One of the more prosaic explanations for these anomalies is that they were the result of observation error. The fact that they have now been seen by both Planck and WMAP eliminates that possibility. Quite what the eventual explanation will be is unclear. The temptation is always to look for the most exotic possible answer, but it may be that, while still very real, the cold spot and the axis of evil might be due to an unexpected foreground and have nothing to do with more fundamental cosmology. The apparent coincidence of the axis of evil with the ecliptic plane of our own Solar System makes a local explanation very tempting, but it is difficult to understand how something so local could produce an effect that looks like a 2.73 K black body.

8.13 PROSPECTS FOR THE FUTURE

To some extent, Planck represents the final word in studies of the intensity variations of the CMB. It measures the strength of all the main features to pretty much as high a precision as you could want (see Figure 8.5). The accuracy of these observations is reflected in the high precision to which the various cosmological parameters are measured. Is this, then, the end of CMB astronomy? Definitely not, because, while Planck might be the final word in measurements of CMB intensity variation, it is almost the first word in the measurement of variations in the polarisation of the CMB.

Any source of electromagnetic radiation can have a degree of polarisation, in addition to having an intensity. Every photon received by a detector has a polarisation resulting from the orientation of the oscillations in the electric and magnetic field that it represents. Many processes, including normal black body radiation, produce photons which have random polarisation orientations, so there is no favoured polarisation direction, and the light can be described as unpolarised. Some processes, however, such as scattering off interstellar dust, can lead to polarisation. A more prosaic example are polarised sun glasses, which only transmit light that is polarised in one direction.

Scattering of CMB photons at the epoch of recombination produce a variety of patterns of polarisation in the CMB. The most interesting of these are the result of gravitational waves, a kind of stretching and compression of space, that interact with the plasma filling the universe at the time of recombination, but which are produced at much earlier times, and associated with the physical processes that drive inflation. The polarisation patterns left by gravitational waves have a unique signature that cannot be mimicked by other scattering effects. But the level of polarisation from them is very small, so they will be hard to measure. Planck should be sensitive enough to set some of the first useful limits on this kind of polarisation in the CMB, but this is much more difficult to extract from the data than a determination of the intensity fluctuations. One reason for this, for example, is that we know relatively little about any contaminating polarised foreground emission.

We are all eagerly awaiting the Planck polarisation results, but there are already other polarisation experiments under way, working on smaller scales, using ground- and balloon-based observations. One experiment, BICEP2, has already claimed a detection of CMB polarisation due to gravitational waves in the early universe [16]. Whether Planck confirms this detection or not[5], it is clear that this is the future of CMB astronomy, and new space missions are already being proposed that will be to polarisation fluctuations what Planck has been to observations of intensity fluctuations.

---

[5]The first analyses of Planck foreground polarisation data suggests that much of the BICEP2 signal might be due to dust in our own galaxy and not gravitational waves in the early universe [134]

8.14   CONCLUSIONS

The CMB can be thought of as the oldest light in the universe, since, before the epoch of recombination, the universe itself was opaque to electromagnetic radiation. As we have seen in this chapter, studies of the CMB, from its initial detection to the age of precision cosmology ushered in by the latest results from Planck, have been central to telling us about the history and nature of the universe we inhabit. While CMB research started off at radio wavelengths with its discovery by Penzias and Wilson, infrared astronomers, working at far-infrared and submm wavelengths, are now central to this work. Bolometer detectors, just like those used in SCUBA and in the SPIRE instrument on Herschel, are what make Planck, and the current generation of ground and balloon-based CMB telescopes, so effective. This technology, and developments such as the TES bolometers working in SCUBA2, will be even more essential for the polarisation studies that are to come, and that are essential for us to understand the earliest stages of the universe.

CHAPTER 9

Future Heat

9.1 THE DAWN OF A NEW MACHINE

Things move quickly when you don't actually have to build anything.

We're at the stage of a new mission that one of my colleagues calls 'the fun bit'.

Things started at a meeting in Pasadena at which some of the first results from the Spitzer mission were presented. As part of the Herschel team I was there, presenting our own plans for what would happen with this mission, at the time still five years from launch.

But planning never stops, and we were all, already, on the lookout for the next project.

A team from the Japanese space agency, JAXA, was also at the Pasadena meeting, presenting plans for a possible future Japanese mission in the far-infrared. They had their own infrared mission, Akari, launching a year later, and were also thinking about what comes next. But their plans were rather more advanced than ours.

There was lots of talking, with the Japanese team being courted by a variety of different groups, all of whom wanted to put an instrument on their future mission. I'm not quite sure how we did it, maybe it was the successful heritage of the joint UK-Japanaese work on Akari, maybe it was the experience we brought, having worked on IRAS, ISO and, now, Herschel, but we turned out to be their favoured set of collaborators to develop an instrument that would go well beyond the capabilities of Herschel.

So now we get to make things a bit more concrete, in terms of both the design of the hardware and the science we'd like to do. Hopefully the capabilities of the instrument that is eventually designed and what is needed from it to do radically new science will come together.

It's a heady experience. We make suggestions for what might be good to look for - molecular hydrogen emission lines from the clouds of gas that formed the first stars just a few hundred million years after the Big Bang. The hardware experts come back with designs that can do that. We suggest the possibility of looking at the broad emission features that come from different types of rock that might be found in planet-forming disks around young stars, and the designs are tweaked again. Can we do both projects with the same instrument? What else can it do?

Unconstrained by cost, and only moderately constrained by the design of the Japanese spacecraft, an instrument concept gradually comes together.

All too soon this freewheeling part of the design process comes to an end and we have to start facing realities. Budgets need to be defined - financial budgets, power budgets, space and data budgets. Proposals need to be written to get funding for more detailed design studies and, eventually, for the instrument to be built. Politics stars to creep in as we build the necessary international collaboration needed to construct such an instrument. And the real world starts to impinge, with funding cuts in both Europe and Japan.

We're told we can't lead the project because of funding constraints, so hand over all that we have done to colleagues elsewhere - a sad but necessary step to keep the project going.

The heady days are over. We're now into the real nitty gritty of a space mission.

As we work through the daily trudge of getting this maturing project to happen, our thoughts turn further to the future. What will come after it? What would we want to do next?

And soon we're back in that small meeting room, at the fun stage of another new mission that would fly even further in the future, if at all.

9.2 WHAT REMAINS TO BE DONE

An astronomer's work is never done, since the more we discover, the more we find there is more to discover. That is simply a truism. Far more justification, alongside concrete goals, technical requirements, designs and costings, are required to get a new astronomical instrumentation project off the ground.

The previous eight chapters have summarised many areas of astrophysics, though by no means all of them, and have highlighted many of the results that infrared astronomy has made possible. This chapter looks in the opposite direction, namely at things we do not currently know, and how we might use infrared astronomy, among other tools, to find them. Infrared astronomy is actually at a very interesting phase at the moment as a generation of infrared instruments, on the ground and in space, have reached, or are reaching, their end. Planck and the Herschel Space Observatory no longer function, since they have run out of the liquid helium needed to keep their instruments working. On the ground, the leading submm telescopes of their time, the James Clerk Maxwell Telescope (JCMT) and the Caltech Submillimetre Observatory are nearing the end of their operational life. At the same time, a new generation of submillimetre telescopes are on the way. The first of these is ALMA, which we touched on at the end of Chapter 7, but more are on the way. Similar developments on the ground are happening in the optical and near-infrared, with a whole new generation of giant telescopes being planned and entering construction.

In space, because it takes a long time to get missions approved, funded and constructed, there is a bit of a pause. A successor to Herschel is planned but not yet fully approved, and the next generation CMB satellite is a few steps behind that. The next great space observatory will be the James Webb Space Telescope, which will do for the near- and mid-infrared what Hubble did for the optical, and the Euclid mission will provide near all-sky coverage for a huge near-infrared survey of galaxies.

But what are all these instruments trying to do?

The great story of infrared astronomy over the last twenty five years has been the exploration of a new waveband, previously unexplored by astronomers. It has given us fundamental insights into processes we knew were underway, but needed to understand better, such as star formation or the late stages of stellar evolution. It has also shown us things that we didn't expect to find, such as dust-dominated galaxies forming stars at an incredible rate, both locally and in the higher redshift universe.

The next phase in the story of infrared astronomy will be somewhat different, with infrared techniques becoming tools for helping us understand bigger problems in astronomy, physics and, hopefully, biology. The key questions for the next decades that infrared astronomy will help answer are among the following:

• How did the very first stars form, and how did their activities lead to the assembly of the first galaxies?

Between the end of recombination, that produced the light of the CMB, and the most distant galaxies that we can see today, there was a time known as the cosmological dark ages. The universe was filled with neutral gas, there were no stars or galaxies, and nothing was generating light. Then something happened. The first stars formed, producing enough emission to ionise the universe. Their lives ended, seeding the space around them with the dust and enriched material we see in the most distant galaxies and quasars. We know these first stars had to exist, but we have never seen one. They would have had to form out of pristine gas, made almost entirely out of hydrogen and helium, so their formation processes would be very different to those of the stars we see forming today. But to see the first stars we need infrared telescopes that are much more sensitive than those currently available.

• How do planetary systems form?

We now have a pretty good idea of the processes behind star formation, as we saw in Chapter 5. We know that planets form at the same time as stars, in the protoplanetary disks around them, but the processes behind that formation are still unclear. This work needs sensitivities and resolving powers that have only just begun to be available in the sub-millimetre with the ALMA telescope, and will need similarly enhanced mid- and near-infrared capabilities.

• What is dark matter?

This question, and the next one, are questions of fundamental physics. The existence of dark matter, and dark energy, have been demonstrated by astrophysics, but are likely to have their explanations in theories of particle physics at the highest energies - energies that might only be reachable by the most powerful particle accelerators available, such as the Large Hadron Collider, and perhaps not even there. Dark matter particles provide the bulk of the gravitational matter in the universe. It is non-baryonic dark matter, rather than the gravitationally insignificant normal, baryonic matter, that produced the large scale structure of the universe and that drives the motions of stars in galaxies, and galaxies in galaxy clusters. We know it is there, and its spatial distribution in (at least) galaxy clusters can be mapped thanks to gravitational lensing. But we do not know what the dark matter is made of. It is likely to be some kind of fundamental particle, but its exact nature currently eludes us. One of the goals of the Euclid mission is to study the nature of dark matter through examining its effects on the growth of structure in the universe.

• What is dark energy?

While we think we know what, in general, the dark matter is, the nature of dark energy is even less clear. The action of dark energy in acceler-ating the expansion rate of the universe can be measured by comparing the brightness and redshift of distant supernovae [142, 127]. But the nature of the dark energy, the origin of what appears to be an outward pressure generated by purely empty space, is far from clear. Unlike dark matter, there isn't even a clear understanding of how dark energy might emerge from theories of fundamental physics[1]. Seeing how dark energy's contribution to the universe changes over time is another goal of the Euclid mission.

• What are the properties of planets around other stars, and do any of them harbour life?

The discovery of exoplanets in the mid-1990s invented a whole new field of astronomy. We now have satellites, like Kepler, dedicated to their discovery. Nearly two thousand exoplanets are known at the time of

---

[1]This isn't to suggest that there are no theories that can explain dark energy. Quite the contrary. Theorists can always come up with an explanation for anything, and usually several. The trick is coming up with an explanation that fits in with everything else we know and that can be tested observationally.

writing[2]. The next step in the new field of exoplanet studies is to find out how these planets work. In our own Solar System we can look at the atmospheres, moons and surfaces of other planets. We have characterised them as terrestrial or gas giants, and can already start to do this in other planetary systems, from measurements of their size and mass. But we want to know much more. Do gas giants in other systems have clouds and giant storms similar to the Great Red Spot on Jupiter? Do they have ring systems? And what are their atmospheres made of? But, above and beyond all of these questions, is one that goes beyond astrophysics. Is there life elsewhere in the universe? The next generations of large telescopes, on the ground and in space, in the infrared or at other wavelengths, may be able to answer this question once and for all.

While these are some of the key questions that are driving the designs of new ground- and space-based telescopes, their work will not be confined to these areas. Many other things remain to be understood and discovered, and the telescopes of the future will be equipped to answer more general questions; whether, for example, they concern physics at the edge of the accretion disk of the supermassive black hole at the centre of our own galaxy, or the detailed chemical and physical processes by which dust condenses out of the ejecta of a supernova, to list but two. And there will always remain the thing that makes astronomy that little bit more exciting than some other fields of science - the unknown and unexpected - which might appear in an observation dedicated to something else entirely.

9.3 SPACE VS. GROUND

The balance between space-based and ground-based telescopes is a delicate one. We have seen that observatories in space can avoid the problems of the Earth's atmosphere, allowing, for example, observations in the far-infrared which are completely impossible from Earth. However, the balance of power between ground- and space-based astronomy is shifting, thanks to a number of technological developments.

One of the key goals of the Hubble Space Telescope (HST), when it was launched in 1990, was to get above the turbulence of the atmosphere rather than its opacity. The HST operates primarily in the optical band, where the atmosphere is transparent. However, that does not mean that the atmosphere has no effect on the incoming radiation. At optical wavelengths, rather than absorbing the incoming light, turbulence in the atmosphere distorts what can be seen through it. The process is similar to what you would see if you were lying at the bottom of a swimming pool looking up through the rippling water at a light overhead. By operating above the atmosphere, the HST was able

---

[2]At the rate exoplanets are being discovered, this number will be well out of date by the time this book is published.

to escape this effect, known as seeing, and produce images of astronomical objects with unrivalled sharpness.

The resolving power of an astronomical instrument fundamentally depends on two things: the size of the main, or primary, mirror, and the wavelength at which it is working. But a telescope beneath the atmosphere is also limited by seeing. The HST has only a 2.5 m main mirror, which means it was a modest-sized telescope even at the time of its launch. Today, we have mirrors on the ground that are four times that size, but their resolving power, the smallest things that they can see, are still limited by atmospheric turbulence.

However, the last twenty five years have seen major progress in adaptive optics, a technology that allows the ripples of the distorting atmosphere to be measured and corrected. One of the key developments necessary for this, the use of lasers to create artificial stars in the upper atmosphere which can be used to measure atmospheric distortions, was originally developed for the US military and was top secret. Scientists who worked with the military and knew about this approach, many of whom were astronomers, had to wait for the technology to be declassified before they could apply it to astronomical observations. That started to happen in the 1990s, and adaptive optics systems are now in place at most major optical telescopes around the world. While somewhat more sophisticated systems are necessary for telescopes significantly larger than 10 m across, this means that the balance of power for high resolution optical and near-infrared observations has shifted from space to the ground. Since you no longer need space to escape atmospheric seeing, and since there are fewer limitations on the size of telescope you can build on the ground, the most powerful next-generation observatories in the optical and near-infrared will be on the ground

A different set of new technologies will lengthen the lives of future space-based infrared telescopes that will operating at wavelengths that cannot penetrate the atmosphere. The main limiting factor for the current observatories is the need to use cryogenic materials, like liquid helium or frozen hydrogen, to keep the detectors and other parts of the instrument at their operating temperatures. These cryogenic materials gradually boil away and, when they are exhausted, the telescope or instrument can no longer function. This is why the Herschel Space Observatory and Planck satellite both ceased operations after about four years in space.

Things have actually been developing in this area for a while, and both Herschel and Planck had lives as long as they did because of this. Herschel, for example, had the same amount of helium coolant on board as the previous European Space Agency infrared mission, ISO (the Infrared Space Observatory). ISO only lasted 18 months before its coolant supplies expired, while Herschel lived more than twice as long. There were two reasons for this. Firstly, Herschel was sent much further away from Earth than ISO, to orbit around a place known as the second Lagrange point, or L2. This is roughly 1.5 million kilometres away from the Earth, and is a place where the gravitational attraction of the Earth and Sun are balanced in such a way that it is easy to keep a spacecraft there. It is also positioned in such a way that the Earth and the Sun are always in the same direction. This allows a satellite like Herschel or Planck to have a large Sun-shield to protect the parts of the spacecraft that need to be kept cold. Things like the primary mirror or the instruments are never heated by direct light from the Sun, or reflected sunlight from the Earth. Instead, they gradually cool down by radiating their heat away into the cold darkness of space. Herschel's primary mirror, for example, cooled to temperatures of about 90 K. This overall scheme is known as passive cooling (as opposed to the active cooling of cryogenic materials) and was devised in the 1990s by infrared astronomers Harley Thronson and Tim Hawarden. Passive cooling is the reason that L2 is now the favourite destination for most astronomical satellites.

Even with passive cooling, satellites like Herschel are still dependent on cryogenics to cool their instruments to operational temperatures that are just a fraction of a degree above absolute zero. While passive cooling means that liquid helium can last much longer, it will still run out in the end. Finding an alternative to liquid helium cooling would help reduce the mass of a satellite, and thus cut launch costs and other limitations, and would, in principle, allow much longer mission lifetimes. This is now becoming possible with the introduction of closed-cycle mechanical coolers that rely on nothing more than electricity, supplied by the satellite's solar panels, to cool instruments to temperatures as low as 0.3 mK. In fact, several elements of the Planck cooling system were mechanical coolers, allowing it to have a mission lifetime comparable to Herschel even though it did not have a large reservoir of liquid helium. Other parts of its cooling system, though, were not closed-cycle, and instead had to use an expendable gas supply which eventually ran out. The next generation of satellites requiring cryogenic temperatures to keep their instruments going are likely to have entirely closed-cycle mechanical cooling systems, allowing them to have much longer operational lifetimes.

9.4 THE EUROPEAN EXTREMELY LARGE TELESCOPE (E-ELT)

To begin our discussion of future astronomical projects relevant to infrared astronomy, we will look at the biggest of them all, the European Extremely Large Telescope (E-ELT). This is a project to build an optical and near-infrared telescope with a primary mirror that is nearly 40 metres across. You can get an idea of the scale of this construction in Figure 9.1. The little blobs on the ground outside the telescope in this picture are cars and trucks. The size of the mirror means that it will, as long as its sophisticated adaptive optics system can correct the atmospheric seeing, be able to produce images nearly a hundred times sharper than those from the HST at the same time as having the power to gather over two hundred times as much light, making it much more sensitive.

While adaptive optics is a key technology allowing E-ELT to produce high resolution images, making a mirror that big in the first place is a difficult

Figure 9.1 Artist's concept of the European Extremely Large Telescope (E-ELT).
With a mirror diameter of 39 m, the E-ELT will be the largest telescope in the world once it is completed. It is currently in the early stages of construction in the Atacama desert in Chile. (Courtesy of ESO; image by L. Calcada http://www.eso.org/public/.)

problem. The solution for this that has been adopted for the E-ELT was first tried on the Keck telescopes. These, at 10 m diameter, are currently among the largest optical telescopes in the world. They use a segmented mirror, which is made up of a series of hexagonal facets that slot together to make the whole. Each of the hexagonal segments has to be made separately, and be figured and polished to be the right shape for their eventual position in the larger mirror. For the Keck telescopes the separate facets are 1.8 m across, with 36 of them combining to make the final mirror. Needless to say, aligning all of these separate segments to make something that functions as a single mirror and not a chaotic mixture of poorly aligned sub-reflectors is not easy. This problem is rather harder for the E-ELT since its primary mirror will be made up of 798 separate segments, each 1.45 m across.

The main science goals for E-ELT include the search for, and study of, the first stars to form in the universe, analysis of the processes by which galaxies assemble in the early universe, and examination of the constituents and behaviour of exoplanet atmospheres. But, as an observatory, users will get to propose whatever science they want to do with its wide range of instruments.

Once its construction is completed in 2022, E-ELT will be the largest telescope in the world. However, it is in fact a scaled down version of the original concept of a large telescope for the 21st century. This was called OWL - the Overwhelmingly Large Telescope [66] - and would have had a primary mirror 100 m across, meaning that it would be able to detect objects 1500 times fainter than the faintest object seen by HST. Sadly, OWL is just too large a project to build at the moment, with a cost in excess of 1.5 billion Euros. One day, though, once E-ELT is a great success, maybe OWL will be built, and astronomers can have access to an optical telescope bigger than the Albert Hall. In the meantime, E-ELT is being planned and constructed by the European Southern Observatory (ESO) in the Atacama desert in Chile, not far from the site of the ESO Very Large Telescope (VLT).

9.5   OTHER GIANT TELESCOPES

Needless to say, the E-ELT is not the only giant optical telescope being planned or built. At least two others are in an advanced stage of planning, with parts of their construction already under way. The first of these, and likely the first of this new generation of enormous telescopes to go into operation, is the Thirty Metre Telescope (TMT). Its primary mirror will be a little smaller than that of the E-ELT, at 30 m in diameter. Like E-ELT and the Kecks before it, the TMT will also use a segmented mirror design, with the final mirror made up of 492 segments, each 1.4 m across. The TMT will be built on Mauna Kea in Hawaii so, unlike the E-ELT in Chile, it will be based in the northern hemisphere. The main science goals for the TMT are broadly similar to those of the E-ELT, and it will feature a range of instruments from the optical to the infrared. The TMT project is led by several US institutions but also has involvement from Canada, Japan, India and China. The first

Figure 9.2 Artist's concepts of the Thirty Metre Telescope (TMT, left) and the Giant Magellan Telescope (GMT, right).
(Courtesy of the TMT Observatory Corporation (left), and Giant Magellan Telescope - GMT Corporation (right).)

steps of its construction are already underway on the summit of Mauna Kea, and an artist's impression of what the completed telescope will look like can be found on the left hand side of Figure 9.2.

The last of these new giants of ground-based optical and near-infrared astronomy is rather different. This is the Giant Magellan Telescope (GMT), an impression of which can be found on the right hand side of Figure 9.2. Rather than taking the segmented mirror approach of E-ELT and TMT, the GMT will combine the light of a smaller number of much larger mirrors to produce a collecting area equivalent to a 22 m diameter telescope and a resolving power equivalent to a 24.5 m telescope. This is done using seven 8.4 m diameter mirrors. These mirrors are similar to those used in most of today's 8 m class telescopes, such as the ESO VLT or Gemini telescopes. The idea of combining a small number of separate circular mirrors to produce a much larger collecting area is not new. The most famous example of this is probably the Multi-Mirror Telescope (MMT). This operated at Mount Hopkins observatory from 1979 to 1998 using a combination of six 1.4 m mirrors, producing the equivalent of a 4.5 m telescope. However, in 1998 the MMT was upgraded to use a single 6.5 m mirror, resulting in it being renamed as the Monolithic Mirror Telescope[3]. A more recent multi-mirror telescope is the Large Binocular Telescope (LBT), which uses two 8.4 m mirrors mounted side by side to provide the equivalent collecting power of a single 11.8 m mirror and the resolving power of a 22.8 m telescope.

With the GMT due for completion in 2020, the TMT in 2018 and the E-ELT in 2022, it is clear that the first years of the next decade will bring revolutionary changes in the capabilities of ground-based telescopes. All of

---

[3]This allowed the name MMT to still be applied to the telescope and was the result of a competition. My favourite option for the new name, though, was the Minor Misnomer Telescope. Sadly this didn't win.

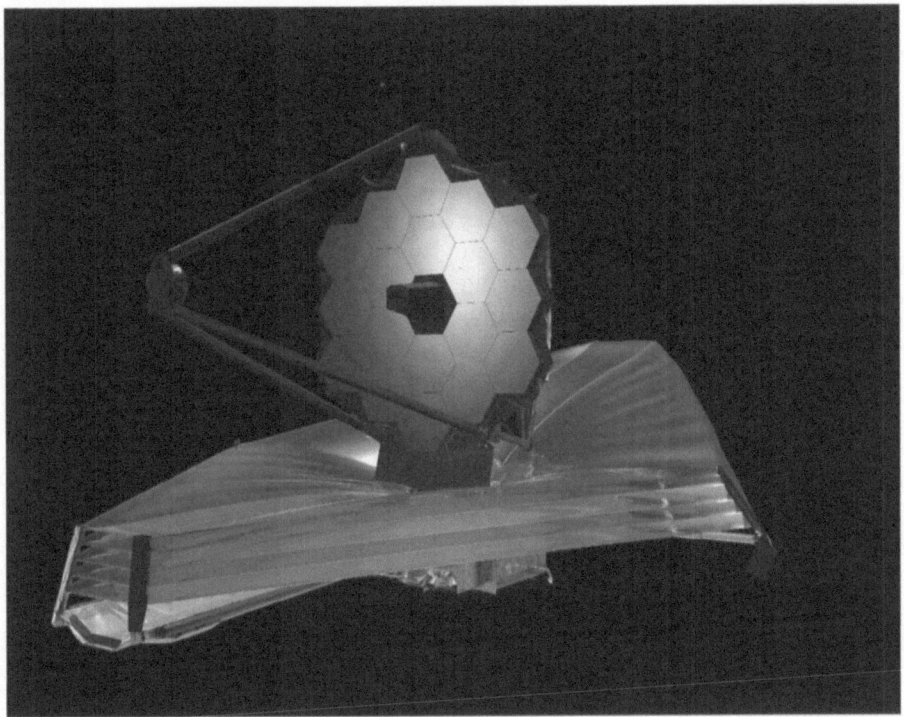

Figure 9.3 Artist's concept of the James Webb Space Telescope. (Courtesy of NASA.)

these mighty new observatories are aimed at searches for the first stars, the study of young galaxies, examination of exoplanets and the other major questions for 21st century astronomy listed at the start of this chapter. And all of them will have excellent capabilities in the near- and mid-infrared since these wavelengths, as we have seen, are central to all of these areas of research.

9.6   THE JAMES WEBB SPACE TELESCOPE (JWST)

While these developments in large ground-based telescopes are on the way, a very powerful new space observatory will also be launched towards the end of the current decade - as long as the huge technical challenges it faces don't lead to further cost increases and delays. This is the James Webb Space Telescope (JWST, see Figure 9.3), the infrared successor to the hugely successful Hubble Space Telescope.

While HST's capabilities were driven by the need to get above atmospheric turbulence, allowing it to observe in the optical and near-ultraviolet with resolutions unachievable from the ground at the time, JWST's capabilities are aimed squarely at the infrared. In fact every instrument on JWST is an infrared instrument, while its capabilities in the optical are restricted to redder wavelengths. The reason for this is that JWST's primary science goals, including the search for the first stars in the universe and the study of planets around other stars, are all better served by infrared than optical observations.

JWST is a very ambitious project and pushes against technological limitations in many areas. First of all, it will have a mirror 6.5 m across, larger than most ground-based telescopes. This is needed to achieve the resolutions necessary to detect the first stars in the universe and to directly image planets around other stars. The problem with this mirror size is that there isn't a space launch vehicle wide enough for it to be launched in one piece. The 3.5 m mirror of Herschel, the largest astronomical mirror in space so far, was limited to that size for this reason. So the JWST mirror is going to be launched in a folded arrangement. Once in space, the mirror will unfold and, hopefully, provide the near perfect optical surface that this telescope requires.

The mirror isn't the only thing that will be launched in a pre-folded configuration. In fact, to be able to accommodate JWST in the launch vehicle, the whole telescope, its vast sunshield and the primary mirror, will all be launched in a folded configuration for eventual self-assembly once it reaches space. All of JWST is on a larger scale than any previous space observatory. The 6.5 m mirror is only the start. For example, the telescope will also be passively cooled to a temperature lower than that of Herschel. This requires a multi-layer sunshield that will be 12.2 m x 18 m in size that has to unfold into its operational configuration.

In space, unless you are very lucky, if something breaks it stays broken. This means that spacecraft engineers are very conservative people, avoiding anything with a tendency to fail unless it is absolutely essential for the mission. Moving parts are anathema, since anything that moves has a chance of getting stuck. Moving parts that are mission critical - meaning that if the part fails the mission fails - are a very scary prospect. The number of mission critical moving parts in JWST is very large. The primary mirror and telescope have to unfold, the multiple layers of the vast Sun shield have to unfurl, the communications antenna and solar panels have to deploy. All are mission critical[4]. If something goes wrong with any of them, then the capabilities of the telescope will be compromised. Many stages of this self deployment are what might be called 'single point failure modes' - if just one doesn't work the mission is dead.

JWST, like most recent space astronomy missions, is going to orbit the L2 point, which lies 1.5 million kilometres away from the Earth. This is necessary for JWST's telescope to passively cool to a temperature of about 40 K, allowing its supremely sensitive operation in the mid-infrared. But this location also means that there is no chance of rescue or repair if something goes wrong. The Hubble Space Telescope, in contrast, is in low Earth orbit, accessible to manned servicing missions. These have been used to constantly update its instruments to the latest technology and, in the first servicing mis-

---

[4]A video of JWSTs deployment is linked to via this book's accompanying webpage.

sion, to correct problems that arose with the shape of its primary mirror. The HST was, in effect, given a pair of spectacles and, since that time, it has been an unalloyed success. No such servicing missions will be possible for JWST, so if the complicated origami deployment goes wrong, there will be no second chance.

The complexity of the spacecraft deployment, and many other factors, have led to long delays and significant increases in cost for JWST. When first approved in 1997, JWST was slated for launch in 2007 and a budget of just half a billion dollars. As the project developed, more reliable figures were produced and, in 2002, as the design was being finalised, a launch date of 2011 and a budget of 2.5 billion dollars seemed reasonable. We are now at the point where the launch has slipped to 2018, and the cost has increased to 8.7 billion dollars. The cost is sufficiently high that it now dwarfs everything else in the NASA space astronomy programme.

In many ways JWST is now the mission that is too big to fail. When it is finally launched, hopefully in 2018, it will not only be engineers and astronomers holding their breath as the mirror, telescope and sunshield unfold, it will also be the many administrators and aerospace company staff who have committed to making it work. When it goes into operation JWST will be an astoundingly powerful instrument, but the story of how it came to be, surviving a ballooning budget and crashing through planned launch dates, will also be astounding.

9.7   THE ATACAMA LARGE MILLIMETRE ARRAY (ALMA)

ALMA, as it is known, was mentioned at the end of Chapter 7 in the context of observations of distant, gravitationally lensed galaxies. It will be one of the future stars of submillimetre astronomy and, despite the fact that it is already taking astronomical data, it is only partially complete. This qualifies it for a mention in this chapter. Unlike all of the telescopes discussed so far, ALMA is an interferometer. Instead of having one mirror, ALMA will be made up of 66 antennae separated by up to 16 kilometres. You can get some idea of the size of the ALMA site from Figure 9.4. The signals from these different antennae are combined to synthesise images that have the resolution of a single telescope 16 kilometres across, giving ALMA resolution at millimetre and submillimetre wavelengths comparable to, and in some circumstances better than, the HST and JWST working in the optical and near-infrared. The 66 antennae, which are mostly 12 m diameter but with some smaller 7 m diameter dishes, also mean that ALMA collects a lot of light, making it very sensitive.

The combination of optical observations with HST or one of the new generation of extremely large ground-based telescopes, mid-infrared observations from JWST and submillimetre interferometry from ALMA will give us detailed, highly sensitive observations across much of the electromagnetic spectrum. This will be hugely effective in understanding everything from the first stars in the universe to the details of exoplanet atmospheres.

Figure 9.4 The ALMA site in Chile.
An aerial view showing both the antennae and some of the roads that allow antennae to be placed up to 16 km apart. (Courtesy of ALMA; http://www.almaobservatory.org/.)

9.8   THE CERRO CHAJNANTOR ATACAMA TELESCOPE (CCAT)

While ALMA will be phenomenally sensitive - able to do in a few hours what would take days or weeks of observation at other telescopes - one weakness is that it doesn't cover large areas of the sky very well. This means it is well suited to detailed studies of objects already known to be of interest that have been found by other telescopes. Among the instruments that will be feeding ALMA targets will be Herschel and the James Clarke Maxwell Telescope, but ALMA on its own will be very slow in tracking down new, interesting targets for detailed study. The Cerro Chajnantor Atacama Telescope (CCAT, a diagram of which can be found in Figure 9.5) will be a new submillimetre telescope, built not far from ALMA, that can survey the sky to depths better than Herschel or JCMT, and provide ALMA with all the interesting sources it might want.

CCAT will have a 25 m mirror, making it the largest single dish submillimetre telescope in the world. It be at an altitude of 5600 m, about 500 m higher than ALMA, and so will observe through skies even more transparent. In the best conditions it is even possible that CCAT will be able to observe at a wavelength of 250 microns, something previously possible only from space. CCAT will be equipped with wide field submillimetre detectors, able to cover large areas of the sky to sensitive limits very quickly. This means that surveys with CCAT will be able to find interesting objects, whether protostars, dusty galaxies, or objects in the outer reaches of our own Solar System, and feed them to ALMA for more detailed study.

Figure 9.5 The design of the CCAT telescope.
(Image used with permission from CCAT; http://www.ccatobservatory.org/.)

9.9 THE SPACE INFRARED TELESCOPE FOR COSMOLOGY AND ASTROPHYSICS (SPICA)

The Space Infrared telescope for Cosmology and Astrophysics (SPICA) is a planned far-infrared space observatory that in concept is a follow-up mission to Herschel. It will have a primary mirror 3 m across, slightly smaller than that of Herschel, and will operate in orbit around the L2 point. One of the key differences between Herschel and SPICA is that SPICA will be kept cold using mechanical coolers rather than expendable liquid helium. This means that SPICA can have a much longer lifetime than Herschel, and will also allow the primary mirror to be actively cooled to much lower temperatures than the passively cooled Herschel mirror. The background radiation falling on the SPICA detectors will be ten thousand times less than the background landing on the Herschel detectors, allowing them to be much more sensitive.

As an observatory, SPICA will be equipped with a range of multipurpose instruments covering a wavelength range from 5 to 200 microns, and maybe longer depending on the final instrument selection. Scientific projects for SPICA include all those listed at the start of this chapter. A particular speciality of SPICA will be opening up the wavelength range between about 25 microns and 60 microns, which is largely unexplored. This wavelength range includes a number of spectral lines that can reveal much about the physical conditions inside regions deeply obscured by dust. SPICA will also have an observing mode that will allow the direct detection of Jupiter-sized planets orbiting other stars.

SPICA was originally a Japanese led project with involvement from the European Space Agency (ESA) and others. However, the funding situation in Japan is so tight that, at the time of writing, it seems as if ESA will have to become a full partner with the Japanese Space Agency (JAXA) for the mission to be viable. This means that involvement in the mission will be substantially more expensive for ESA, and makes its future somewhat uncertain. If it goes ahead, SPICA is likely to be launched in the late 2020s.

9.10 EUCLID

Most of the projects described so far in this chapter have been observatories, designed and capable of making a wide variety of observations. While their capabilities are defined by a set of science goals that, at the time of planning and proposal, are likely to be at the centre of their operations, they can be used for any number of other projects. The next couple of projects, both space missions, are rather more focused in their intent. They are being designed to ask some very specific questions through their observations. However, those same observations will also be useful for other purposes bringing what is termed 'legacy' value to these missions.

The Euclid Mission (see Figure 9.6) is an ideal example of this kind of project. The central aim of this mission is to better understand the roles of

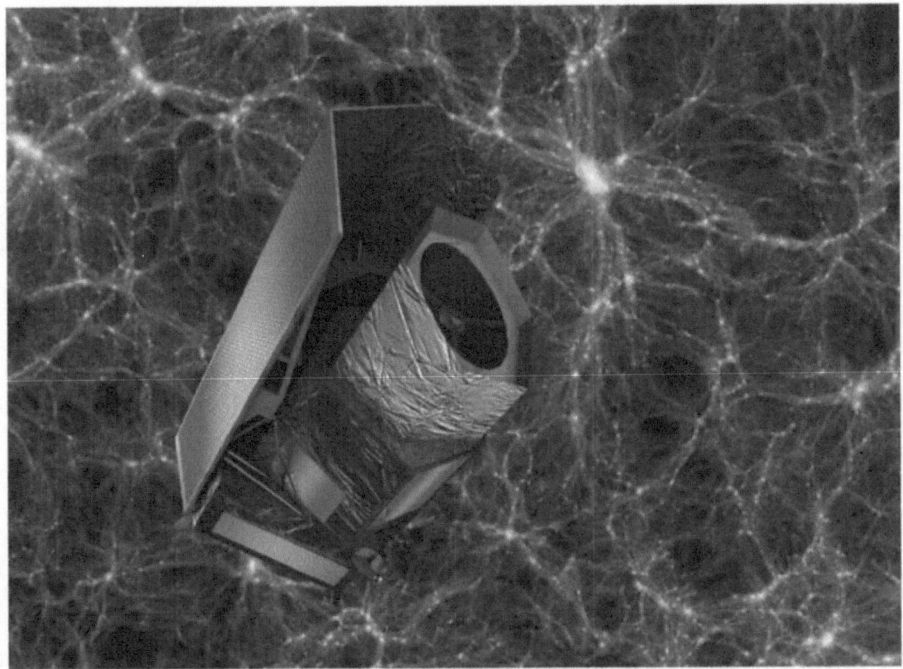

Figure 9.6 Artist's concept of the Euclid satellite.
This image shows the Euclid satellite against the cosmic web of dark matter that defines the large scale structure of galaxies in the universe. Observing the effects of the dark matter making up this cosmic web is one of the central goals of the mission. (Courtesy of ESA; http://www.esa.int/ESA.)

dark matter and dark energy in the universe. It will do this by measuring the distribution of mass in the universe through looking at both the distribution of galaxies, and the way mass distorts the images of these galaxies through weak gravitational lensing. The observations Euclid needs to achieve these goals are essentially a very large imaging survey of the extragalactic sky, using a combination of optical and near-infrared wavelengths, in conjunction with complementary ground based observations and some near-infrared spectroscopy. The end result will be multicolour images of over a billion galaxies at resolutions comparable to those of the HST, as well as spectroscopic data for some fraction of these sources.

This is a vast amount of data, all of which must be processed, analysed and combined to produce the final results dealing with dark matter and dark energy. The dark energy result, for example, could amount to just a few numbers, describing how the contribution of dark energy to the universe changes over time. This level of data compression is rather like that of the Planck project, where maps of the entire sky at nine different frequencies were boiled down to just a few numbers describing the underlying cosmological parameters of the universe.

However, the vast archive of data that Euclid will produce can be used for many other purposes. The headline results from Euclid on dark matter and dark energy may change, or confirm, our views of fundamental physics, but there will be much more astrophysics that can be done with the legacy archive.

9.11   CORE/PRISM

You can always tell when a project is in the early stages of development when it has yet to settle on a name for itself. This is the case for CoRE/PRISM, short for Cosmic Origins Explorer/ Polarised Radiation and Spectroscopy Mission, a project which is essentially a proposal for the next generation CMB satellite after Planck. The aim of this project is to look at the polarisation of the CMB with far greater accuracy than was possible with Planck, with the intent of detecting the fingerprints of cosmic inflation and the physics behind it. CoRE/PRISM is currently being planned as a proposal for the fourth ESA medium sized mission (known as M4). The eventual design is still to be determined, but it is clear that this mission will require many more detectors and produce much more data than Planck. While the primary goals for the mission, like those for Euclid and Planck, will be to calculate a small number of numbers that define the nature of cosmic inflation, the huge volume of data that CoRE/PRISM will also produce will provide a vast resource of legacy data for use in many other areas of astronomy.

9.12   ASTRONOMY AS DATA SCIENCE

Euclid and CoRE/PRISM are examples of how astrophysics is moving from an observatory oriented science, where astronomers propose a set of observations of their favoured objects, get the data, then analyse it themselves, to a data oriented science, where vast databases of observations already performed can be queried, searched and analysed. Observatory science will still be part of the field, since detailed observations of interesting sources using exquisitely powerful instruments like JWST, ALMA or E-ELT, will still be needed. But alongside this more traditional approach astronomers will become data scientists using the vast legacy databases.

These databases will come not only from satellites, like Euclid and CoRE/PRISM, but also from ground-based projects which survey large areas of the sky. Some of these projects are already underway, like the VISTA near-infrared and Pan-STARRS optical surveys, while others, like the Sloan Digital Sky Survey (SDSS) have already completed much of their work. All of these projects have, or will, produce a vast quantity of data which will be made public. SDSS, for example, has produced images for 230 million stars and galaxies. The data volume is so large that no single person can deal with all this data. Instead it is processed and analysed automatically, and the results are made available over the internet for everyone, professional or amateur, to use. Access to this data has spawned a number of citizen science projects, such as Galaxy Zoo [98] where members of the public can view galaxy images and classify them as spirals or ellipticals, and look for unusual objects. This has produced a number of interesting results (e.g., [99]), and the value of citizen scientists for such projects is clear.

However, the next generation of giant surveys will be too big for the army of citizen scientists gathered by SDSS to study. Euclid will produce detailed images of over a billion galaxies, for example, while the biggest legacy survey project of them all, the Large Synoptic Survey Telescope (LSST, see Figure 9.7), will go even further. LSST is a planned 8 m class optical telescope that will observe the whole southern sky - a bit over half the entire sky - in six different colours once every five nights for at least a decade. One of its central aims is to look for objects whose light output varies, but this data can also be combined to produce very sensitive images, or some mix of the two to allow the variability of fainter objects to be studied. The LSST will produce about 30 terabytes of data each night.

The databases to be produced by Euclid and LSST will be so large that only a tiny fraction of them can ever be examined in detail by human beings. All the effort of the Galaxy Zoo volunteers over the years would only be capable of dealing with a few percent of these datasets - useful, certainly, but far from complete. Instead, the new data scientist astronomers will use techniques such as machine learning and robust statistics to sift through the catalogs and images to make new discoveries.

Figure 9.7 An artist's rendition of the Large Synoptic Survey Telescope. (Image used with permission from the LSST Corporation.)

9.13   OTHER WAVELENGTHS

Other parts of the electromagnetic spectrum will also be examined by the next generation of observatories on the ground and in space. Among such projects will be two major new facilities at opposite ends of the spectrum - a new, powerful X-ray telescope in space, currently called Athena, and a vast new array of radio telescopes that will form the Square Kilometre Array (SKA).

The Athena satellite has been selected as the next ESA large mission, to be launched in the 2020s. Athena's central aims are to look at the physics of black holes, how they grow and how they shape the universe, and to examine how normal matter assembles into the large scale structures we see in the universe today. Observations at X-ray wavelengths allow matter at temperatures of millions of degrees to be studied, something that cannot be done at other wavelengths.

At the very other end of the spectrum, in the radio, SKA (Figure 9.8) is among the most exciting of the new observatories. With elements in South Africa and Australia, and including hundreds of thousands of separate radio receivers, it will be able to both produce images 50 times sharper than HST, and cover large areas of the sky. It represents a vast leap in the capabilities of radio astronomers. SKA will be used for a wide range of studies, including searches for the clouds of hydrogen that collapsed to form the first stars, studies of cosmic magnetic fields, and observations of the Earth's own ionosphere. SKA will even be sensitive enough to detect air traffic control radars run by any extraterrestrial intelligences that might exist on planets closer than 50 light years away.

Astronomy is also extending beyond the electromagnetic spectrum to search for different types of signal coming from the cosmos. This work includes the search for gravitational waves - distortions in the fabric of space that are produced when large masses, like neutron stars or black holes, collide with each other (e.g., [42, 75]) . Meanwhile, high energy neutrinos are being searched for using cubic kilometres of ice at the South Pole. These very weakly interacting particles, produced by energetic astrophysical processes such as black holes, supernovae and the early universe, will occasionally hit an atomic nucleus in the ice, producing a flash of light. These flashes are detected by strings of photomultipliers hung deep in the antarctic ice sheet. The first results from this project, appropriately named ICECUBE, have recently been released [82].

Where once the infrared was the wild frontier of astronomy, the next few decades will take us to wavelengths and types of radiation that were undreamed of in William Herschel's time.

9.14   CONCLUSIONS

In this book we have seen how infrared astronomy has provided insights into the workings of the universe that are unavailable at other wavelengths. This

Figure 9.8 An artist's rendition of part of the Square Kilometre Array. Two different types of SKA antenna are visible here: the dishes of the survey telescope in the upper half of the image and the antennas of the low frequency array, to the right of the car. (Image used with permission from the SKA Organisation; https://www.skatelescope.org.)

applies on all scales, whether in our own Solar System, looking at star formation in our own Galaxy, or looking at the formation of the entire universe. When William Herschel discovered infrared light in 1800 he could not have imagined the impact it would have, whether on his own subject of astronomy or in the broader world. Today, rescue workers use the same kind of detectors we use for mid-infrared astronomy to search for earthquake victims, far-infrared light is used to check aircraft passengers for hidden weapons, and Herschel's kind of infrared light, just outside the range of what we can see with our own eyes, is used in television remote controls. None of this would have happened without Herschel's initial insight that led to the discovery of the infrared, or without the astronomers, physicists and engineers who have laboured long and hard to uncover the secrets of the universe hidden at wavelengths beyond our optical gaze.

Infrared astronomy is now at the heart of the subject, used by all astronomers to study their chosen targets. This book is but a brief summary of what they have achieved so far. We have come a long way, but, as we have seen in this chapter, there is a lot more to be done.

CHAPTER 10

Further Reading

This book has presented a rather rapid overview of many areas of science and technology associated with infrared astronomy. If you would like to read more about any of these areas in detail, I recommend the following books and other resources as a place to start:

10.1   CHAPTER 1: FINDING THE HEAT

Night Vision: Exploring the Infrared Universe, by Michael Rowan-Robinson, published by the Cambridge University Press

Introducing Particle Physics: A Graphic Guide, by Tom Whyntie, published by Icon Books

Introducing Quantum Theory: A Graphic Guide, by J.P. McEvoy, published by Icon Books

In Search of Schrodinger's Cat, by John Gribbin, published by Black Swan

10.2   CHAPTER 2: FEELING THE HEAT

Astronomical Observations: An Optical Perspective, by Gordon Walker, published by the Cambridge University Press

Observational Astronomy: Techniques and Instrumentation, by Edward Sutton, published by the Cambridge University Press

Detection of Light: from the ultraviolet to the submillimetre, by G.H. Rieke, published by Cambridge University Press

Astronomy with Charge Coupled Devices, by R.A. Jansen, e-book, http://www.public.asu.edu/~rjansen/ast598/ast598_jansen2006.pdf

Tools of Radio Astronomy, by T.L. Wilson, K. Rohlfs, & S. Hüttermeister, published by Springer

A movie about the BLAST balloon telescope is available. See `http://www.devlinpix.com/film/blast`

10.3   CHAPTER 3: LOCAL HEAT

An Introduction to the Solar System, edited by David A. Rothery, Neil McBride & Iain Gilmour, published by the Cambridge University Press

Discovering the Solar System, by Barrie W. Jones, published by Wiley

Exploring the Solar System, by Peter Bond, published by Wiley-Blackwell

10.4   CHAPTER 4: THE HEAT OF THE STARS

An Introduction to the Sun and Stars, edited by Simon F. Greem & Mark H. Jones, published by the Cambridge University Press

The Stars: their structure and evolution, by R.J. Tayler, published by the Cambridge University Press

An Introduction to the Theory of Stellar Structure and Evolution, by Dina Prialnik, published by the Cambridge university Press

10.5   CHAPTER 5: YOUNG HEAT

An Introduction to the Sun and Stars, edited by Simon F. Greem & Mark H. Jones, published by the Cambridge University Press

The Formation of Stars, by Steven W. Stahler & Francesco Palla, published by Wiley

An Introduction to Star Formation, by Derek Ward-Thompson and Anthony P. Whitworth, published by the Cambridge University Press

10.6   CHAPTER 6: DISTANT HEAT

Night Vision: Exploring the Infrared Universe, by Michael Rowan-Robinson, published by the Cambridge University Press

Galaxies: A Very Short Introduction, by John Gribbin, published by the Oxford University Press

Active Galactic Nuclei, by Julian H. Krolik, published by the Princeton University Press

Galactic Astronomy, by James Binney, published by the Princeton University Press

Revealing the Heart of the Galaxy, by Robert H. Sanders, published by the Cambridge University Press

Gravity's Fatal Attraction, by Mitchell Begelman & Martin Rees, published by the Cambridge University Press

The Physics & Evolution of Active Galactic Nuclei, by Hagai Netzer, published by the Cambridge University Press

Galaxy: Mapping the Cosmos, by James Geach, published by Reaktion Books

10.7   CHAPTER 7: ANCIENT HEAT

Galaxy Formation and Evolution, by Houjon Mo, Frank van den Bosch, & Simon White, published by the Cambridge University Press

Galaxy Formation and Evolution, by Hyron Spinrad, published by Springer

Observational Cosmology, by Stephen Serjeant, published by the Cambridge University Press

Gravity's Fatal Attraction, by Mitchell Begelman & Martin Rees, published by the Cambridge University Press

The Physics & Evolution of Active Galactic Nuclei, by Hagai Netzer, published by the Cambridge University Press

Galaxy: Mapping the Cosmos, by James Geach, published by Reaktion Books

10.8   CHAPTER 8: FIRST HEAT

Observational Cosmology, by Stephen Serjeant, published by the Cambridge University Press

Wrinkles in Time, by George Smoot & Keay Davidson, published by Harper Perennial

The Very First Light, by John Boslough, published by Basic Books

The First Three Minutes, by Stephen Weinberg, published by Basic Books

The Inflationary Universe, by Alan Guth, published by Basic Books

Just Six Numbers, by Martin Rees, published by Basic Books

Cosmological Physics, by John Peacock, published by the Cambridge University Press

10.9   CHAPTER 9: FUTURE HEAT

The Last of the Great Observatories, by George H. Rieke, published by the University of Arizona Press

The Design and Construction of Large Optical Telescopes, edited by Pierre Bely, published by Springer

Giant Telescopes: Astronomical Ambition and the Promise of Technology, by W. Patrick Mccray, published by the Harvard University Press

Bibliography

[1] André P. The Herschel View of Star Formation. *ArXiv e-prints*, 1309.7762, September 2013.

[2] André P. et al. Submillimeter continuum observations of Rho Ophiuchi A - The candidate protostar VLA 1623 and prestellar clumps. *ApJ*, 406:122–141, March 1993.

[3] André P. Observations of Protostars and Protostellar Stages. *The Cold Universe*, eds. T. Montmerle, C.J.Lada, I.F.Mirabel, & J.Tran Thanh Van, Editions Frontiéres, 1994.

[4] André P. et al. From filamentary clouds to prestellar cores to the stellar IMF: Initial highlights from the Herschel Gould Belt Survey. *A&A*, 518:L102, July 2010.

[5] Antonucci R. R. J. & Miller J. S. Spectropolarimetry and the nature of NGC 1068. *ApJ*, 297:621–632, October 1985.

[6] Apai D. et al. HST Spectral Mapping of L/T Transition Brown Dwarfs Reveals Cloud Thickness Variations. *ApJ*, 768:121, May 2013.

[7] Arnaboldi M. et al. ESO VISTA Public Surveys - A Status Overview. *The Messenger*, 149:7–11, September 2012.

[8] Arp H. Atlas of Peculiar Galaxies. *ApJS*, 14:1, November 1966.

[9] Auger M. W. et al. The Sloan Lens ACS Survey. X. Stellar, Dynamical, and Total Mass Correlations of Massive Early-type Galaxies. *ApJ*, 724:511–525, November 2010.

[10] Aumann H. H. et al. Discovery of a shell around Alpha Lyrae. *ApJ.*, 278:L23–L27, March 1984.

[11] Banerjee S. et al. The emergence of super-canonical stars in R136-type starburst clusters. *MNRAS*, 426:1416–1426, October 2012.

[12] Barnes J. E. & Hernquist L. E. Fueling starburst galaxies with gas-rich mergers. *ApJL*, 370:L65–L68, April 1991.

[13] Becklin E. E. & Zuckerman B. A low-temperature companion to a white dwarf star. *Nature*, 336:656–658, December 1988.

[14] Beichman C et al. The Coldest Brown Dwarf (or Free-floating Planet)?: The Y Dwarf WISE 1828+2650. *ApJ*, 764:101, February 2013.

[15] Bennett A. S. The preparation of the revised 3C catalogue of radio sources. *MNRAS*, 125:75, 1962.

[16] BICEP2 Collaboration et al. BICEP2 I: Detection Of B-mode Polarization at Degree Angular Scales. *ArXiv e-prints*, March 2014.

[17] Bindschadler D. L. Magellan: A new view of Venus' geology and geophysics. *Reviews of Geophysics*, 33:459–468, 1995.

[18] Block D. L. et al. An almost head-on collision as the origin of two off-centre rings in the Andromeda galaxy. *Nature*, 443:832–834, October 2006.

[19] Boley A.C. et al. Constraining the Planetary System of Fomalhaut Using High-resolution ALMA Observations. *ApJL*, 750:L21, May 2012.

[20] Borucki W. J. et al. Kepler Planet-Detection Mission: Introduction and First Results. *Science*, 327:977, February 2010.

[21] Boulanger F. & Perault M. et al. Diffuse infrared emission from the galaxy. I - Solar neighborhood. *ApJ.*, 330:964–985, July 1988.

[22] Bouwens R. J. et al. A candidate redshift z∼10 galaxy and rapid changes in that population at an age of 500Myr. *Nature*, 469:504–507, January 2011.

[23] Bouwens R. J. et al. UV-Continuum Slopes of >4000 z∼4-8 Galaxies from the HUDF/XDF, HUDF09, ERS, CANDELS-South, and CANDELS-North Fields. *ArXiv e-prints*, June 2013.

[24] Boyanovsky D. et al. Phase Transitions in the Early and Present Universe. *Annual Review of Nuclear and Particle Science*, 56:441–500, November 2006.

[25] Brammer G. B. et al. A Tentative Detection of an Emission Line at 1.6 micron for the z ∼ 12 Candidate UDFj-39546284. *ApJL*, 765:L2, March 2013.

[26] Brown M. E. et al. Discovery of a Planetary-sized Object in the Scattered Kuiper Belt. *ApJL*, 635:L97–L100, December 2005.

[27] Canup R. M. Origin of Saturn's rings and inner moons by mass removal from a lost Titan-sized satellite. *Nature*, 468:943–946, December 2010.

[28] Chapman C. R. et al. Surface properties of asteroids - A synthesis of polarimetry, radiometry, and spectrophotometry. *Icarus*, 25:104–130, May 1975.

[29] Chapman S. C. et al. A Redshift Survey of the Submillimeter Galaxy Population. *ApJ*, 622:772–796, April 2005.

[30] Charbonneau D. et al. Detection of an Extrasolar Planet Atmosphere. *ApJ*, 568:377–384, March 2002.

[31] Clements D. L. et al. Optical imaging of ultraluminous IRAS galaxies: how many are mergers? *MNRAS*, 279:477–497, March 1996.

[32] Clements D. L. & Couch W. J. Candidate primeval galaxies in the Hubble Deep Field. *MNRAS*, 280:L43–L47, June 1996.

[33] Clements D. L. et al. Candidate high-redshift and primeval galaxies in Hubble Deep Field South. *MNRAS*, 308:L11–L16, September 1999.

[34] Clements D. L. et al. The SCUBA Half-Degree Extragalactic Survey (SHADES) - VIII. The nature of faint submillimetre galaxies in SHADES, SWIRE and SXDF surveys. *MNRAS*, 387:247–267, June 2008.

[35] Collier Cameron A. et al. WASP-1b and WASP-2b: two new transiting exoplanets detected with SuperWASP and SOPHIE. *MNRAS*, 375:951–957, March 2007.

[36] Coppin K. et al. The SCUBA Half-Degree Extragalactic Survey - II. Submillimetre maps, catalogue and number counts. *MNRAS*, 372:1621–1652, November 2006.

[37] Coraddu M. et al. Deuterium burning in Jupiter interior. *Physica A Statistical Mechanics and its Applications*, 305:282–286, March 2002.

[38] Corey B. E. & Wilkinson D. T. A Measurement of the Cosmic Microwave Background Anisotropy at 19 GHz. In *Bulletin of the American Astronomical Society*, volume 8, page 351, March 1976.

[39] Crowther P.A. et al. The R136 star cluster hosts several stars whose individual masses greatly exceed the accepted $150M_{solar}$ stellar mass limit. *MNRAS*, 408:731–751, October 2010.

[40] Cruz M. et al. The CMB cold spot: texture, cluster or void? *MNRAS*, 390:913–919, November 2008.

[41] Dainty J.C. & Shaw R. *Image Science*. Academic Press, 1974.

[42] Danzmann K. et al. The Proposed Evolved-LISA Mission. In *AAS/High Energy Astrophysics Division*, volume 13 of *AAS/High Energy Astrophysics Division*, page 302.01, April 2013.

[43] Das S. et al. The Atacama Cosmology Telescope: A Measurement of the Cosmic Microwave Background Power Spectrum at 148 and 218 GHz from the 2008 Southern Survey. *ApJ*, 729:62, March 2011.

[44] Davidson K. & Humphreys R. M. Eta Carinae and Its Environment. *ARAA*, 35:1–32, 1997.

[45] Decin L. et al. Warm water vapour in the sooty outflow from a luminous carbon star. *Nature*, 467:64–67, September 2010.

[46] Dole H. et al. The cosmic infrared background resolved by Spitzer. Contributions of mid-infrared galaxies to the far-infrared background. *A&A*, 451:417–429, May 2006.

[47] Dougherty M. K. et al. Identification of a Dynamic Atmosphere at Enceladus with the Cassini Magnetometer. *Science*, 311:1406–1409, March 2006.

[48] Dowell C.D. et al. HerMES: Candidate High-Redshift Galaxies Discovered with Herschel/SPIRE. *ApJ*, 780:75, January 2014.

[49] Downes D. et al. Proposed identification of Hubble Deep Field submillimeter source HDF 850.1. *A&A*, 347:809–820, July 1999.

[50] Eales S. et al. The Canada-UK Deep Submillimeter Survey. IV. The Survey of the 14 Hour Field. *AJ*, 120:2244–2268, November 2000.

[51] Eales S. et al. The Herschel ATLAS. *PASP*, 122:499–515, May 2010.

[52] Eiroa C. et al. DUNES: DUst around NEarby Stars. A Herschel Open Time Key Programme. In M. Chavez, E. Bertone, O. Vega, and V. De la Luz, editors, *New Quests in Stellar Astrophysics III: A Panchromatic View of Solar-Like Stars, With and Without Planets*, volume 472 of *Astronomical Society of the Pacific Conference Series*, page 23, April 2013.

[53] Ellis R. S. et al. The Abundance of Star-forming Galaxies in the Redshift Range 8.5-12: New Results from the 2012 Hubble Ultra Deep Field Campaign. *ApJL*, 763:L7, January 2013.

[54] Eriksen H. K. et al. Asymmetries in the Cosmic Microwave Background Anisotropy Field. *ApJ*, 605:14–20, April 2004.

[55] Evans J. E. & Maunder E.E. Experiments as to the actuality of the 'canals' observed on Mars. *MNRAS*, 63:488, 1903.

[56] Faure C. et al. First Catalog of Strong Lens Candidates in the COSMOS Field. *ApJS*, 176:19–38, May 2008.

[57] Fischer J. et al. Herschel-PACS spectroscopic diagnostics of local ULIRGs: Conditions and kinematics in Markarian 231. *A&A*, 518:L41, July 2010.

[58] Fixsen D. J. et al. The Spectrum of the Extragalactic Far-Infrared Background from the COBE FIRAS Observations. *ApJ*, 508:123–128, November 1998.

[59] Flint, A. *Popular Science Monthly*, 45, 1894.

[60] Fonti S. & Marzo G. A. Mapping the methane on Mars. *A&A*, 512:A51, March 2010.

[61] Ford G. P. et al. Herschel Exploitation of Local Galaxy Andromeda (HELGA). III. The Star Formation Law in M31. *ApJ.*, 769:55, May 2013.

[62] Gahm G.F. et al. Mass and motion of globulettes in the Rosette Nebula. *A&A*, 555:A57, July 2013.

[63] García-Hernández D.A. et al. Formation of Fullerenes in H-containing Planetary Nebulae. *ApJL*, 724:L39–L43, November 2010.

[64] Gillessen S. et al. Monitoring Stellar Orbits Around the Massive Black Hole in the Galactic Center. *ApJ*, 692:1075–1109, February 2009.

[65] Gillessen S. et al. Pericenter Passage of the Gas Cloud G2 in the Galactic Center. *ApJ*, 774:44, September 2013.

[66] Gilmozzi R. & Dierickx P. OWL concept study. *The Messenger*, 100:1–10, June 2000.

[67] Greaves J.S. et al. A Dust Ring around epsilon Eridani: Analog to the Young Solar System. *ApJL*, 506:L133–L137, October 1998.

[68] Green S.F. & Jones M.H. *An Introduction to the Sun and Stars*. Cambridge University Press, 2004.

[69] Greco J. P. et al. Measurement of the Mass and Stellar Population Distribution in M82 with the LBT. *ApJ*, 757:24, September 2012.

[70] Griffin M.J. et al. The Herschel-SPIRE instrument and its in-flight performance. *A&A*, 518:L3, July 2010.

[71] Griffith C. A. et al. Titan's surface and troposphere, investigated with ground-based, near-infrared observations. *Icarus*, 93:362–378, October 1991.

[72] Griffith C. A. et al. The evolution of Titan's mid-latitude clouds. *Science*, 310(5747):474–477, 2005.

[73] Grill-Spector K. & Malach R. The human visual cortex. *Annual Reviews Neuroscience*, 27:649, 2004.

[74] Hanany S. et al. MAXIMA-1: A Measurement of the Cosmic Microwave Background Anisotropy on Angular Scales of 10' to 5 degrees. *ApJL*, 545:L5–L9, December 2000.

[75] Harry G. M. et al. Advanced LIGO: the next generation of gravitational wave detectors. *Classical and Quantum Gravity*, 27(8):084006, April 2010.

[76] Henry P. S. Isotropy of the 3 K Background. *Nature*, 231:516–518, June 1971.

[77] Holland W. S. et al. SCUBA: a common-user submillimetre camera operating on the James Clerk Maxwell Telescope. *MNRAS*, 303:659–672, March 1999.

[78] Holland W.S. et al. Scuba-2: the 10 000 pixel bolometer camera on the James Clerk Maxwell Telescope. *MNRAS*, 430:2513–2533, April 2013.

[79] Holmes R. *The Age of Wonder, HarperPress*. January 2008.

[80] Huggins W. & Huggins M. L. Note on the Photographic Spectrum of the Great Nebula in Orion. *Sidereal Messenger, vol. 9, pp.314-318*, 9:314–318, August 1890.

[81] Hughes D. H. et al. High-redshift star formation in the Hubble Deep Field revealed by a submillimetre-wavelength survey. *Nature*, 394:241–247, July 1998.

[82] IceCube Collaboration et al. Searches for High-energy Neutrino Emission in the Galaxy with the Combined ICECUBE-AMANDA Detector. *ApJ*, 763:33, January 2013.

[83] Illingworth G. D. et al. The HST eXtreme Deep Field (XDF): Combining All ACS and WFC3/IR Data on the HUDF Region into the Deepest Field Ever. *ApJS*, 209:6, November 2013.

[84] Inada N. et al. The Sloan Digital Sky Survey Quasar Lens Search. V. Final Catalog from the Seventh Data Release. *AJ*, 143:119, May 2012.

[85] Iping R. C. et al. Detection of a Hot Binary Companion of η Carinae. *ApJL*, 633:L37–L40, November 2005.

[86] Ivison R. J. et al. The SCUBA HAlf Degree Extragalactic Survey - III. Identification of radio and mid-infrared counterparts to submillimetre galaxies. *MNRAS*, 380:199–228, September 2007.

[87] Jameson K.A. et al. Richer colour experience in observers with multiple photopigment opsin genes. *Psychonomic Bulletin and Review*, 8:244, 2001.

[88] Keene J. et al. Spitzer Observations of Centaurus A. In *American Astronomical Society Meeting Abstracts #204*, volume 36 of *Bulletin of the American Astronomical Society*, page 725, May 2004.

[89] Keisler R. et al. A Measurement of the Damping Tail of the Cosmic Microwave Background Power Spectrum with the South Pole Telescope. *ApJ*, 743:28, December 2011.

[90] Kirkpatrick J.D. 2MASS Data Mining and the M, L, and T Dwarf Archives. In E. Martín, editor, *Brown Dwarfs*, volume 211 of *IAU Symposium*, page 189, June 2003.

[91] Könyves V. et al. The Aquila prestellar core population revealed by Herschel. *A&A*, 518:L106, July 2010.

[92] Kroupa P. On the variation of the initial mass function. *MNRAS*, 322:231–246, April 2001.

[93] Kroupa P. The initial mass function of simple and composite stellar populations. arXiv:astro-ph/0703124, March 2007.

[94] Küppers M et al. Localized sources of water vapour on the dwarf planet (1)Ceres. *Nature*, 505:525–527, January 2014.

[95] Latham D.W. The effects of push development and hydrogen sensitization on the detective performance of Kodak plates types iia and 127-04. *AAS Phot. Bull.*, 13:9, 1976.

[96] Lawrence A. Blank-field submm sources, failed stars and the dark matter. *MNRAS*, 323:147–158, May 2001.

[97] Leggett S. et al. Cool White Dwarfs Found in the UKIRT Infrared Deep Sky Survey. *ApJ*, 735:62, July 2011.

[98] Lintott C. J. et al. Galaxy Zoo: morphologies derived from visual inspection of galaxies from the Sloan Digital Sky Survey. *MNRAS*, 389:1179–1189, September 2008.

[99] Lintott C. J. et al. Galaxy Zoo: 'Hanny's Voorwerp', a quasar light echo? *MNRAS*, 399:129–140, October 2009.

[100] Liu M.C. et al. The Extremely Red, Young L Dwarf PSO J318-22: A Free-Floating Planetary-Mass Analog to Directly Imaged Young Gas-Giant Planets. *ArXiv e-prints*, October 2013.

[101] Lonsdale C. et al. First Insights into the Spitzer Wide-Area Infrared Extragalactic Legacy Survey (SWIRE) Galaxy Populations. *ApJS*, 154:54–59, September 2004.

[102] Luhman K.L. Discovery of a Binary Brown Dwarf at 2 pc from the Sun. *ApJLett*, 767:L1, April 2013.

[103] Lunine J. I. & Atreya S. K. The methane cycle on Titan. *Nature Geoscience*, 1:159–164, March 2008.

[104] Magorrian J. et al. The Demography of Massive Dark Objects in Galaxy Centers. *AJ*, 115:2285–2305, June 1998.

[105] Martin C.D. et al. The galaxy evolution explorer: A space ultraviolet survey mission. *ApJ*, 619:L1, 2005.

[106] Mauskopf P. D. et al. Measurement of a Peak in the Cosmic Microwave Background Power Spectrum from the North American Test Flight of Boomerang. *ApJL*, 536:L59–L62, June 2000.

[107] Mawet D et al. The Vector Vortex Coronagraph: Laboratory Results and First Light at Palomar Observatory. *ApJ*, 709:53–57, January 2010.

[108] Mayor M. & Queloz D. A Jupiter-mass companion to a solar-type star. *Nature*, 378:355–359, November 1995.

[109] McConnell N.J. et al. Two ten-billion-solar-mass black holes at the centres of giant elliptical galaxies. *Nature*, 480:215–218, December 2011.

[110] McNutt R. et al. Enceladus Gas Plume: Composition and Structure. In *39th COSPAR Scientific Assembly*, volume 39 of *COSPAR Meeting*, page 1221, July 2012.

[111] Meadows V. S. et al. The Spitzer First Look Survey-Ecliptic Plane Component: Asteroids and Zodiacal Background. *ApJS*, 154:469–474, September 2004.

[112] Mennesson B et al. An Interferometric Study of the Fomalhaut Inner Debris Disk. II. Keck Nuller Mid-infrared Observations. *ApJ*, 763:119, February 2013.

[113] Moore J. M. Meteorology: Practical and Applied, 2nd Ed. Published by Rebman Ltd., London, p.297, fig.86, 1910

[114] Mortier A. M. J. et al. The SCUBA Half-Degree Extragalactic Survey - I. Survey motivation, design and data processing. *MNRAS*, 363:563–580, October 2005.

[115] Mumma M. J. et al. Strong Release of Methane on Mars in Northern Summer 2003. *Science*, 323:1041–, February 2009.

[116] Myers S. T. et al. The Cosmic Lens All-Sky Survey - I. Source selection and observations. *MNRAS*, 341:1–12, May 2003.

[117] Nakajima T. et al. Discovery of a cool brown dwarf. *Nature*, 378:463–465, November 1995.

[118] Negrello M. et al. The Detection of a Population of Submillimeter-Bright, Strongly Lensed Galaxies. *Science*, 330:800–, November 2010.

[119] Nelan E. P. et al. Resolving OB Systems in the Carina Nebula with the Hubble Space Telescope Fine Guidance Sensor. *AJ*, 128:323–329, July 2004.

[120] Netterfield C. B. et al. A Measurement of the Angular Power Spectrum of the Anisotropy in the Cosmic Microwave Background. *ApJ*, 474:47, January 1997.

[121] Nolta M. R. et al. The MAT/TOCO Measurement of the Angular Power Spectrum of the Cosmic Microwave Background at 30 and 40 GHz. *ApJ*, 598:97–101, November 2003.

[122] Oesch P. A. et al. The Bright End of the Ultraviolet Luminosity Function at $z\sim 8$: New Constraints from CANDELS Data in GOODS-South. *ApJ*, 759:135, November 2012.

[123] Oesch P. A. et al. Probing the Dawn of Galaxies at $z\sim 9$-12: New Constraints from HUDF12/XDF and CANDELS data. *ApJ*, 773:75, August 2013.

[124] Oke J. B. & Gunn J. E. The Distance of BL Lacertae. *ApJ*, 189:L5, April 1974.

[125] Oliver S.J et al. The Herschel Multi-tiered Extragalactic Survey: HerMES. *MNRAS*, 424:1614–1635, August 2012.

[126] Peacock J. A. *Cosmological Physics, Cambridge University Press*. January 1999.

[127] Perlmutter S. et al. Measurements of Omega and Lambda from 42 High-Redshift Supernovae. *ApJ*, 517:565–586, June 1999.

[128] Petigura E. A. et al. Prevalence of Earth-size planets orbiting Sun-like stars. *Proceedings of the National Academy of Science*, 110:19273–19278, November 2013.

[129] Planck Collaboration et al. Planck 2013 results. I. Overview of products and scientific results. *ArXiv e-prints*, March 2013.

[130] Planck Collaboration et al. Planck 2013 results. XII. Component separation. *ArXiv e-prints*, March 2013.

[131] Planck Collaboration et al. Planck 2013 results. XIII. Galactic CO emission. *ArXiv e-prints*, March 2013.

[132] Planck Collaboration et al. Planck 2013 results. XV. CMB power spectra and likelihood. *ArXiv e-prints*, March 2013.

[133] Planck Collaboration et al. Planck 2013 results. XVI. Cosmological parameters. *ArXiv e-prints*, March 2013.

[134] Planck Collaboration et al. Planck intermediate results. XXX. The angular power spectrum of polarized dust emission at intermediate and high Galactic latitudes. *ArXiv e-prints*, September 2014.

[135] Polletta M. et al. Spectral Energy Distributions of Hard X-Ray Selected Active Galactic Nuclei in the XMM-Newton Medium Deep Survey. *ApJ*, 663:81–102, July 2007.

[136] Preibisch T. et al. Herschel far-infrared observations of the Carina Nebula complex. I. Introduction and global cloud structure. *A&A*, 541:A132, May 2012.

[137] Puget J.-L. et al. Tentative detection of a cosmic far-infrared background with COBE. *A&A*, 308:L5, April 1996.

[138] Radomski J. T. et al. Gemini Imaging of Mid-Infrared Emission from the Nuclear Region of Centaurus A. *ApJ*, 681:141–150, July 2008.

[139] Rebolo R. et al. Discovery of a brown dwarf in the Pleiades star cluster. *Nature*, 377:129–131, September 1995.

[140] Rees M. J. Appearance of Relativistically Expanding Radio Sources. *Nature*, 211:468–470, July 1966.

[141] Riechers D. A. et al. A dust-obscured massive maximum-starburst galaxy at a redshift of 6.34. *Nature*, 496:329–333, April 2013.

[142] Riess A. G. et al. Observational Evidence from Supernovae for an Accelerating Universe and a Cosmological Constant. *AJ*, 116:1009–1038, September 1998.

[143] Roe H. G. et al. Geographic control of titan's mid-latitude clouds. *Science*, 310(5747):477–479, 2005.

[144] Rossetto B.M. et al. The dark energy survey: Prospects for resolved stellar populations. *AJ*, 141:185, June 2011.

[145] Rowan-Robinson M. *Night Vision*. Cambridge University Press, May 2013.

[146] Sanders D. B. et al. Ultraluminous infrared galaxies and the origin of quasars. *ApJ*, 325:74–91, February 1988.

[147] Sanders D. B. & Mirabel I. F. Luminous Infrared Galaxies. *ARAA*, 34:749, 1996.

[148] Schaffer K. K. et al. The First Public Release of South Pole Telescope Data: Maps of a 95 deg² Field from 2008 Observations. *ApJ*, 743:90, December 2011.

[149] Schmidt M. 3C 273 : A Star-Like Object with Large Red-Shift. *Nature*, 197:1040, March 1963.

[150] Schröder K.-P. & Connon Smith R. Distant future of the Sun and Earth revisited. *MNRAS*, 386:155–163, May 2008.

[151] Scott R.L. et al. The use of forming gas in hypersensitizing Kodak spectroscopic plates. *AAS Phot. Bull.*, 15:12, 1977.

[152] Seyfert C. K. Nuclear Emission in Spiral Nebulae. *ApJ*, 97:28, January 1943.

[153] Silk J. Cosmic Black-Body Radiation and Galaxy Formation. *ApJ*, 151:459, February 1968.

[154] Skrutskie M.F. et al. The Two Micron All Sky Survey (2MASS). *AJ*, 131:1163–1183, February 2006.

[155] Smith M. W. L. et al. The Herschel Exploitation of Local Galaxy Andromeda (HELGA). II. Dust and Gas in Andromeda. *ApJ.*, 756:40, September 2012.

[156] Smoot G. F. et al. Detection of anisotropy in the cosmic blackbody radiation. *Physical Review Letters*, 39:898–901, October 1977.

[157] Stahler S. W. & Palla F. *The Formation of Stars Wiley*. January 2005.

[158] Stauffer J.R. et al. Near- and Mid-Infrared Photometry of the Pleiades and a New List of Substellar Candidate Members. *ApJS*, 172:663–685, October 2007.

[159] Sumi T. et al. Unbound or distant planetary mass population detected by gravitational microlensing. *Nature*, 473:349–352, May 2011.

[160] Sykes M. V. et al. The discovery of dust trails in the orbits of periodic comets. *Science*, 232:1115–1117, May 1986.

[161] Tedesco E. F. & Desert F.-X. The Infrared Space Observatory Deep Asteroid Search. *AJ*, 123:2070–2082, April 2002.

[162] Tedesco E. F. et al. The Supplemental IRAS Minor Planet Survey. *AJ*, 123:1056–1085, February 2002.

[163] Treu T. Strong Lensing by Galaxies. *ARAA*, 48:87–125, September 2010.

[164] Urry, C.M., & Padovani, P. Unified Schemes for Radio-Loud Active Galactic Nuclei. *PASP*, 107:803, 1995.

[165] Usui F. et al. AKARI/IRC Mid-Infrared Asteroid Survey. *LPI Contributions*, 1667:6119, May 2012.

[166] van der Marel R. P. et al. The M31 Velocity Vector. III. Future Milky Way M31-M33 Orbital Evolution, Merging, and Fate of the Sun. *ApJ*, 753:9, July 2012.

[167] Van Meeteren A. On the detective quantum efficiency of the human eye. *Vision Research*, 18:257, 1978.

[168] Vidal-Madjar A. et al. An extended upper atmosphere around the extrasolar planet HD209458b. *Nature*, 422:143–146, March 2003.

[169] Vieira J. D. et al., Dusty starburst galaxies in the early Universe as revealed by gravitational lensing. *Nature*, 495:344–347, March 2013.

[170] Walker R. G. et al. Observations of comet IRAS-Araki-Alcock 1983d. *ApJ.*, 278:L11–L14, March 1984.

[171] Walter F. et al. The intense starburst HDF 850.1 in a galaxy overdensity at z=5.2 in the Hubble Deep Field. *Nature*, 486:233–236, June 2012.

[172] Warren S.J. et al. A very cool brown dwarf in UKIDSS DR1. *MNRAS*, 381:1400–1412, November 2007.

[173] Warren S.J. et al. The United Kingdom Infrared Telescope Infrared Deep Sky Survey First Data Release. *MNRAS*, 375:213–226, February 2007.

[174] Welty A. D. Discovery of a Pre-Main-Sequence Spectrscopic Binary: V773 Tauri. *AJ*, 110:776, August 1995.

[175] Whitney A. R. et al. Quasars Revisited: Rapid Time Variations Observed Via Very-Long-Baseline Interferometry. *Science*, 173:225–230, July 1971.

[176] Williams R. E. et al. The Hubble Deep Field: Observations, Data Reduction, and Galaxy Photometry. *AJ*, 112:1335, October 1996.

[177] Wright E.L. et al. The Wide-field Infrared Survey Explorer (WISE): Mission Description and Initial On-orbit Performance. *AJ*, 140:1868–1881, December 2010.

Index